Causal Space-Time

Emergent Physics from Causality

Second Edition

Richard D. Bateson

Cover Photo:
Joueurs de cartes by Georges Mestrallet.
La photographie aurait été prise à Goncelin au café Hustache. Les personnes représentées sont :
- à gauche : Célestin Mangournet
- au centre : André Roux, berger d'Arles
- à droite : M. Billon, marbrier à Goncelin
© Musée Dauphinois, Grenoble, France.

ISBN 9781456377557

drrichardbateson@gmail.com

Preface to the Second Edition

In 2010 I published the first edition of Causal Space-Time whilst at the London Centre for Nanotechnology (LCN), University College London (UCL). At the LCN my research interests diverged from nanotechnology, spin-ice magnetic monopoles and two-dimensional magnetism into the arcane world of causality and its role in quantum mechanics. My discovery that the simplest causal network and based loosely on Reichenbach's principle of common cause, could exactly model the Dirac equation, and be consistent with both special relativity and quantum mechanical statistics, lead me to consider the role of causality in physics. Is it possible that the "elementary" particles and "fundamental" laws of physics are emergent from causality and probability? Many historical philosophers such as Kant would have perhaps taken this for granted but today causality is unfortunately an unfashionable subject in physics. In this book we briefly explore these ideas. I apologise in advance to many readers for the mathematics in Chapter 2 (but this is a physics book after all) and you can skip to Chapter 3 for a summary of the results.

The ideas in the first edition were well received at the EmerQuM 11: Emergent Quantum Mechanics 2011 (Heinz von Foerster Congress) conference in Vienna. A paper was subsequently published, *A causal net approach to relativistic quantum mechanics**, in the Journal of Physics (IOP). In this second edition, the paper is included as an Appendix. Also included is a recent extension of the previous work entitled *A causal net approach to gravitation and dark matter,* which considers quantum causal nets in curved space-time and as a model for gravitation. The new paper outlines how causal nets connect to Einstein's general relativity and in cosmology can describe a "flat" universe. It turns out that curvature of space-time in general relativity can be represented as a variation of the density of possible events comprising the causal net. Lastly, the paper provides a speculative calculation for the ratio of matter to dark energy in the universe that agrees with experimentally observed values to within 1%.

Richard Bateson
Cavendish Laboratory,
Cambridge,
February 2023

* R D Bateson 2012 J. Phys.: Conf. Ser. 361 012009

Contents

Glossary of Notations

The most commonly used notations in alphabetical order with non-latin symbols at the end. Some have several uses but are distinguishable from their context.

c	speed of light
E	energy
G	Newton's gravitational constant
h	Planck's constant
m	mass
p	momentum
$\mathcal{P}$	probability
$\hat{\boldsymbol{p}}$	momentum operator
s	space-time interval
S	action or reference frame
t	time
V	potential
x	space coordinate
v	velocity

$\hat{\boldsymbol{\sigma}}$	vector of Pauli matrices
$\boldsymbol{\sigma}_k$	Pauli matrix ($k = 1,2,3$)
ψ	spinor component or wavefunction
ϕ	probability amplitude
λ	helicity
θ	net angle
τ	proper time
$\boldsymbol{\Psi}$	Dirac spinor
γ	Lorentz factor

Chapter 1

Introduction

1.1 Introduction.

In this book we shall unashamedly attempt to construct a theory of physics based on the fundamental concepts of causality and probability. We shall discover that with just these two ideas we can, from first principles, provide a theory that is compatible with both Einstein's special relativity and quantum mechanics. From a causal network of "events" in space-time we can derive exactly the famous Dirac equation that governs the evolution of the quantum wavefunction. We postulate that fermions along with the phenomenon of spin are "emergent" quasiparticles of the causal net in a similar fashion to well-known quasiparticles in solid-state physics such as phonons and holes in semiconductors. In the low velocity limit the causal net is consistent with the Schrödinger equation and extending to different momentum states we can demonstrate consistency with the Feynman path integral approach to quantum mechanics that allows calculation of well known quantum phenomena such as diffraction.

This Chapter provides the background to understand this theory including the notions of causality, quantum mechanics and emergence. The mathematical causal space-time theory is presented in Chapter 2 but for those readers who dislike mathematics the results are summarised and discussed in Chapter 3.

1.2 Causality.

Causality, the concept of cause and effect, has long been viewed by philosophers as a cornerstone of our understanding of the universe. Often it is considered a notion that must in some way precede physics since it is required in its experience and subsequent understanding. The definition of causality and its role in nature has developed throughout history and unfortunately today remains very much the domain of philosophers, having long since been abandoned by the majority of the physics community. In the recent comprehensive book about the nature of time and the universe, *From Eternity to Here*, by well known cosmologist Sean Carroll the mention of causality is limited to a few lines. This is in my view a disappointing reflection of modern physics, that a topic as fundamental as causality lies largely ignored and is seen as pretty much irrelevant. As physicists we are taught to think in terms of forces and potentials rather than abstract cause and effect. It is pretty simple and black and white for physicists – if an object is accelerating it is being acted on a by a force (a cause) and if it is moving in a straight line there is no force (so no cause). This is however a bit of a throw-back to Newtonian mechanics and the indoctrination of a modern physics education and unfortunately ignores the rich history and philosophy of causality.

1.3 A Brief History of Causality.

1.3.1   *The Ancient Greeks.*

Plato first stated a principle of causality as "everything that becomes or changes must do so owing to some cause; for nothing can come to be without a cause". Aristotle went further and discussed a theory of causation in his *Posterior Analytics* with a pretty wide ranging definition of causality where he defined four type of cause (*aitia*) to describe something: the material cause, the formal cause, the efficient cause and the final cause.

- The *material cause* being the "raw material" or constituents from which something is produced.
- The *formal cause* is what something is intended or planned to be.
- The *efficient cause* is the external influence or entity from which a change starts.
- The *final cause* is the purpose for which something is meant to serve or the purpose for which something exists.

Only really the efficient cause is what we now associate with causation. In addition Aristotle introduced the notions of proper and accidental causation and potential or actual causation.

1.3.2   *The Stoics.*

Stoicism was a highly durable school of Hellenistic philosophy founded by Zeno of Citium in Athens in the 3rd century B.C. It was widely practiced during the Roman Empire and into the 5th Century A.D. until it was banned by the Emperor Justinian 1st – who decided it was not consistent with the Christian world-view. Any major school of thought that dominated the world for so long and included some very clever people such as Cato the younger, Seneca (Fig. 1.1), Epictetus and Marcus Aurelius must have made some sense to educated, rational people living in such a brutal era. Stoicism principally taught a methodology on how to live using logic, ethics and exerting self-control over negative emotions and situations seeming dictated by fate and fortune.

Figure 1.1: The Stoic philosopher Seneca who was ordered to kill himself by Nero when falsely accused of a plot against him. © CORBIS

The Stoics believed that the cosmos was a living, divine organism and that every event in the universe is ordained by fate. They believed that every event in the universe has a cause and is brought about by certain causal conditions. Stoics thus had a sort of *principle of universal causation* – maintaining that no uncaused events could exist and with causal connections existing between all events. In the words of one Stoic author: "Prior events are causes of those following them, and in this manner all things are bound together with one another, and thus nothing happens in the world such that something else is not entirely a consequence of it and attached to it as a cause...From everything that happens something else follows depending on it by necessity as a cause."

Importantly, the Stoics had a view of causality that is very close to our current western outlook.

1.3.3   *The Rationalists and Hume.*

The rationalist philosophers (Descartes, Hobbes, Spinosa and Leibniz) believed that causal inference is an *a priori* concept – that is causation is an innate concept that is required to interpret any experience in nature. Hume disagreed and insisted that by experience of many instances of causation we have an association of ideas that leads us to understand and interpret causation. For Hume causation was definitely not *a priori* to our understanding of the universe. It is interesting to briefly review the views on causality that these philosophers held.

Descartes (1596–1650) broke with tradition and rejected Aristotle's four types of cause. He retained only the efficient cause and divided efficient causes into two categories: (i) particular causes which are the laws of nature and (ii) one general cause which is God who maintains the quantity of motion in the universe.

The English philosopher Thomas Hobbes (1588–1679) retained Aristotle's efficient and material causes but defined causation in terms of relations between moving bodies: "the aggregate of accidents in the agent or agents, requisite for the production of the effect". He proposed that all causation occurs by contact between bodies and not at a distance. In the universe according to Hobbes "all the effects that have been, or shall be produced, have their necessity in things antecedent" and all causal relations are determined by God.

The Dutchman, Spinoza (1632–1677) emphasised the necessity between cause and effect: "From a given determinate cause an effect necessarily follows; and, on the other hand, if no determinate cause can be given it is impossible that an effect can follow." Spinoza rejected the idea of final causation and held that necessary causes are all based on a causal, mathematical order from other causes with God being the only free cause.

Gottfried Leibniz (1646-1716), who we will talk about more later, was, as well as a philosopher, a pretty amazing inventor and mathematician. He invented a calculating machine called the Stepped Reckoner which he demonstrated in 1670 to the Royal Society– the first machine that could perform the four basic mathematical functions! As a mathematician he is accredited with discovering calculus independently from Newton. His philosophical arguments were wide ranging. In his *Theodicee*, ridiculed by Voltaire in *Candide*, he argued that despite the imperfections in the world it must be the "best amongst all possible worlds". His views on causality deviated in an original way from his predecessors. Although he maintained "there is nothing without a reason, or no effect without a cause", he introduced the concept of "monads" as the ultimate constituents of reality. In an attempt to preserve Aristotle's formal and final causation he proposed that each monad has its own purpose and causal independence but develops in a harmonious synchrony with the other monads. Efficient causes exist in a preordained way towards final causes in a manner pre-determined by God.

David Hume (1711–1776) claimed that although we think that causation involves the idea of *necessity* – with causes necessarily preceding effects – this is not true. To Hume causal relationships we establish are illusionary and simply impressions we have from expectation, habit and empirical observation. Thus apparent causal necessity is not a logical necessity we can formally derive. In doing so he refuted the ideas of the Rationalists before him.

1.3.4 *Kant.*

Immanuel Kant (1724–1804) was motivated by a strong disagreement with the views on causality expressed by Hume and saw several flaws in Hume's arguments. The Leibnizian theory of "innate" ideas – concepts that are presupposed in any experience and are *a priori* concepts was elaborated by Kant [1]. In his famous *Transcendental Deduction* Kant recognised the importance of causal connections – "the relation of cause and effect is the condition of the objective validity of our empirical judgement" and "causality leads to the concept of action, this in turn to the concept of force, and thereby to the concept of substance". Kant argued that Leibniz's *principle of sufficient reason* (Section 1.4.1) becomes the law of causality – the law that every event in the empirical world is bound by causal connections. Space and time for Kant were not "substances" but instead elements of a framework for structuring experience, so that spatial and temporal measurements are used to define the distance between causal events.

Figure 1.2: Immanuel Kant argued causality was an *a priori* concept and logically preceded other concepts in nature. © CORBIS

1.3.5   *Newtonian causality.*

Newton, in his masterpiece *Philosophiae Naturalis Principia Mathematica* (1687), detailed his three famous laws of motion that were clearly stated in causal terms:

1. Every body perseveres in its state of rest, or of uniform motion in a right line, unless it is compelled to change that state by forces impressed thereon.
2. The alteration of motion is ever proportional to the motive force impressed; and is made in the direction of the right line in which that force is impressed.

3. To every action there is always opposed an equal reaction; or the mutual actions of two bodies upon each other are always equal, and directed to contrary parts.

These laws can be briefly summarised as: (i) a body continues in uniform motion unless acted on by a force, (ii) forces provide acceleration and (iii) to every force there is an equivalent reaction. Thus in some sense with Newton's laws forces become essentially causes. Since if a body in uniform motion is not acted on by a force (and not accelerating) the first law of motion is an uncaused event. Thus to some philosophers Newton rejected the law of universal causation – that every event must have a cause – a philosophical ideology that had historically prevailed. Others go further and claim that under Newton events can happen either according to law or those that are the effects of causes and the two cases are mutually exclusive.

1.3.6   *A physicist's definition of causality.*

After all these various somewhat confusing philosopher's views of causality we shall seek refuge in a well-defined physicist's definition of causality. Max Born in 1949 came up with the following assumptions which provide a workable definition.

1. Causality postulates that there are laws by which the occurrence of an entity B of a certain class depends on the occurrence of an entity A of another class, where the word entity means any physical object, phenomenon, situation or event. A is called the cause, B the effect.
2. Antecedence postulates that the cause must be prior to, or at least simultaneous with, the effect.
3. Contiguity postulates that cause and effect must be in spatial contact or connected by a chain of intermediate things in contact.

Various people have argued that such a definition is redundant with the discovery of quantum mechanics. However, as we shall see in Chapter 2 this definition actually provides a good, if not exact, starting point for our definition of a causal net which models quantum mechanical statistics. Care must however be taken to ensure that any definition of simultaneity of cause and effect in clause (2) is compatible with the principles of relativity and we shall discuss this later in Section 1.7.4.

1.4 Absolute and Relative Space-Time.

1.4.1   *The Leibniz–Clarke correspondence.*

Newtonian mechanics is consistent with the concept of absolute space–time. The movement of all objects is with respect to a fixed background of space and time. A coordinate system where time and space are orthogonal can be provided and all things measured with respect to this (Fig. 1.3). Absolute space can be viewed as a perfect 3 dimensional box in which objects move.

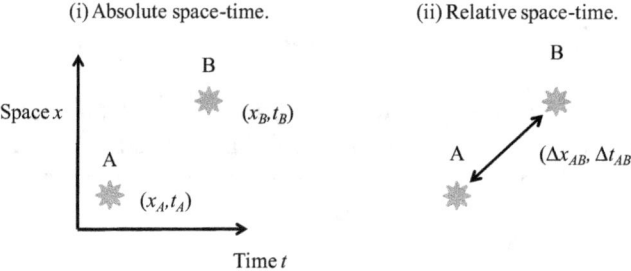

Figure 1.3: Absolute and relative space-time for events A and B.

Interestingly, Newton himself did not really claim that space and time were absolute and left open the possibility of space and time being defined by the relative distance between objects. Newton's ardent supporters were however much more adamant that space-time is absolute and not relative. Champion amongst these was Clarke who entered into a heated debate with the principal proponent of relative space-time, the heavy weight German philosopher and scientist Leibniz. This exchange of letters forms the well-known Leibniz–Clarke correspondence [2]. Most of the reasoning followed by both parties now seems rather dated and obscure. Much recourse is made to invoking the role of God in deciding whether space-time is relative or absolute and Leibniz makes much use of his favourite line of reasoning which is the principle of sufficient reason – that "There must be a sufficient reason [often known only to God] for anything to exist, for any event to occur, for any truth to obtain." Leibniz's views on relative space-time are succinctly expressed in his third letter: "For my part, I have stated more than once that I hold space to be something purely relative, as time is: that I hold it to be an order of coexistences, as time is an order of successions. For space denotes in terms of possibility and ordering of things which exist at the same time, in so far as they exist together, without considering their particular ways of existing."

Unfortunately, Leibniz died before the end of the correspondence. It is probably the case that the success of Newtonian mechanics lead to the general perception that space-time was absolute for the next couple of hundred years despite the serious objections of Berkley (1685–1753) and Mach (1838–1916).

1.4.2 *Einstein's special relativity.*

In 1905 Einstein proposed two axioms which lead to the *principle of relativity* and a paradigm shift in the way we think about space and time [3]. The first axiom concerned the interdependence of physical laws from the choice of inertial system (different frames of movement) and the second axiom constancy of the speed of light.

1. The principle of relativity – the laws of physics are identical in all inertial frames.
2. The principle of invariant light speed – there exists an inertial frame in which light signals in a vacuum always travel rectilinearly at constant speed c, in all directions, independently of the motion of the source.

Both of these statements by themselves seem pretty reasonable but combined they lead to the astonishing result that light signals are propagated with the same speed in *all* inertial frames. The notion that however fast we move the speed of light is the same is contrary to our classical concept of space and time. Mathematically it provides the invariant space-time interval Δs which relates intervals of space x and time t between two reference frames say S and S'

$$\Delta s^2 = (c\Delta t')^2 - (\Delta x')^2 = (c\Delta t)^2 - (\Delta x)^2 \qquad (1.1)$$

and the well-known Lorentz transforms between the space and time coordinates in both frames. Importantly, relativity meant that defining simultaneity of events between different inertial frames became impossible although causality must be preserved.

Special relativity dispenses with the notion of absolute space-time and the existence of a luminiferous ether through which electromagnetic and other waves would propagate. The constancy of the speed of light was supported by the famous Michelson–Morley experiment. However, the strange relativistic effects predicted such as time dilation, Lorentz contraction and composition of velocities are totally counter-intuitive. These effects have proven readily experimentally verifiable and even the widely used GPS navigation system includes a relativistic correction.

1.5 Quantum Phenomena.

Quantisation in nature was really first observed with the discovery of the photo-electric effect – the emission of electrons from the surface of metals when illuminated by light. Einstein provided an explanation that the energy in a beam of light is carried in discrete packets called quanta or photons. People were used to thinking of light in terms of waves but this was the first glimpse of two new quantum phenomena – quantisation of energy and *wave-particle duality*. Quantisation of energy was also apparent in the quantum energy levels and spectra of electron orbitals in atoms. Wave-particle duality of matter was further developed by de Broglie (Nobel prize 1929) [4] and wave diffraction has been regularly observed in slit experiments with numerous particles from electrons to relatively massive carbon buckyballs. A new theory, pioneered by Einstein, Bohr, Schrödinger and others, to describe these strange quantum phenomena at microscopic scales was developed – quantum mechanics.

1.6 The Quantum Mechanics.

1.6.1 *The quantum wavefunction.*

In quantum mechanics the most fundamental concept is the wavefunction ψ which encapsulates all the information of a quantum system. Its modulus-squared $|\psi|^2 = \psi^*\psi$ represents a probability – a positive, real number [5]. For a particle this represents the probability of finding a particle at a particular location r at time t and is thus the probability distribution $P(r, t) = |\psi(r, t)|^2$. Importantly a wavefunction is comprised of an amplitude and a phase – which varies in space and time like an actual wave. In general by computing the probability we lose information about the phase. In many experimental setups and quantum calculations the effect of the phase is apparent in interference effects between different wavefunctions, for example in slit diffraction. The simplest and most important wavefunction is the plane-wave or free particle wavefunction – not acted on by any forces or potentials – which has uniform probability across infinite space and time although the phase varies. One might think that this is an abstract special case but in fact by combining plane-wave states with different energy (or momentum) we can mathematically create any arbitrary wavefunction. The wavefunction itself is totally deterministic and describes the probabilities of finding a particle for all points in time and space – thus although a particle has an undefined position its probability distribution is well specified.

1.6.2   *The Schrödinger equation.*

The well-known equation that governs the development of the wavefunction is the Schrödinger equation [5]. This is a strange second order differential equation that provides a solution to the wavefunction ψ for a given set of boundary conditions such as energy potentials. The solutions can be represented as a sum or superposition of plane-wave states. It is extraordinarily successful fundamental equation and is used in a wide range of theoretical and practical applications. The equation is notoriously confusing when first encountered since it looks like a cross between a complex diffusion equation for the wavefunction with strange remnants of Newtonian mechanics thrown in. The Schrödinger equation is a non-relativistic equation, for particles with velocities well below the speed of light, and can be used to describe a wide range of quantum particles.

1.6.3   *The Dirac equation.*

In an attempt to provide a relativistic form of the Schrödinger equation the brilliant Paul Dirac (Fig. 1.4) in 1928 came up with the Dirac equation [6]. The Dirac equation is a truly amazing equation and is *relativistically invariant* which means it is compatible with special relativity in the limit of particle velocities approaching the speed of light. In the limit of low velocities it can be easily shown to reduce to the Schrödinger equation, so in a sense the Schrödinger equation is a low velocity approximation of the Dirac equation in the same way Newtonian mechanics is a low velocity form of Einstein's special relativity. Not surprisingly Dirac was very proud of his equation and after he discovered it he was slightly afraid of rigorously testing it in case he proved it wrong! At first glance the Dirac equation is very complicated, even when compared to the Schrodinger equation. It is composed of several 4x4 matrices and the wavefunctions are 4-component wavevectors. The reason for this is that in making the Schrödinger equation fully relativistic Dirac discovered two new concepts – particle *spin* and negative energy states. The Dirac equation actually apparently contained 4 solutions comprising 2 spin states and 2 energy states.

Figure 1.4: The brilliant Paul Dirac. On meeting the flamboyant Richard Feynman for the first time Dirac apparently asked "I have an equation do you have one too?" © CORBIS

The basic Dirac equation is found to describe a particle with 2 spin states corresponding to a spin $\pm\frac{1}{2}$ particle. Such a particle is termed a *fermion* (non-integral spin) rather than a boson (spin 0 or integral spin). The simplest and commonest fermion we encounter is the well known electron. Every fundamental particle is a fermion (there are 12 of them known to date including the leptons and quarks) (see Table 1.1) and other particles are formed from these to make composite particles, nuclei and atoms. Nobody really knows what electron spin really is – it is probably not a particle actually spinning – but it is a property of the particle that can be aligned along different spatial directions and can interact with other particles, fields and potentials. For example, the neutron is a *composite* fermion called a hadron and is composed of 3 fermions (1 up quark and 2 down quarks) providing a net spin $\pm\frac{1}{2}$. It is widely used as a probe of magnetism, at places such as ISIS (Oxford) and ILL (Grenoble), since it can be readily polarised into different spin states which interact with magnetic fields in materials.

Table 1.1: The 12 known fermions (excluding antiparticles).

Fermion		Family	Mass (MeV)	Charge (e)
electron neutrino	ν_e	lepton	<0.0022	0
muon neutrino	ν_μ	lepton	<0.17	0
tau neutrino	ν_τ	lepton	<15.5	0
electron	e	lepton	0.511	-1
muon	μ	lepton	105.7	-1
tau	τ	lepton	1774.87	-1
up	u	quark	2.4	2/3
down	d	quark	4.8	-1/3
charm	c	quark	1270	2/3
strange	s	quark	104	-1/3
top	t	quark	171,200	2/3
bottom	b	quark	4,200	-1/3

The second set of solutions to the Dirac equation correspond to negative energy states. Now these were initially also very confusing and not very well understood. Dirac was none too happy with them and thought that, since a higher energy state could always transition to a lower negative energy state by radiating away the energy, these represented a major problem. He invented his famous negative "electron sea" or Dirac sea of filled negative energy states to avoid this. He predicted that there must be an anti-particle to the electron and initially Dirac thought the negative states corresponded to the proton. Unfortunately, the proton mass was far too big relative to the electron but his theory was rescued by the discovery of the *positron* in 1932. A modern interpretation in quantum field theory (QFT) is that negative energy solutions really do correspond to antiparticles and serious quantum calculations require the Dirac equation to include these negative energy states. Several specialised phenomena that actually rely on the existence of negative energy states (such as the Klein tunnelling paradox) have been predicted but to my knowledge none have been experimentally demonstrated. It is hoped that forthcoming experiments on graphene (layers of carbon with atomic thickness) will demonstrate these relativistic negative energy tunnelling effects.

1.6.4   *Dirac and quantum operators.*

The mathematical formalism of quantum mechanics has been developed around the concept of operators. For example, the mathematical operator for momentum $\hat{p}$ "operates" on a wavefunction and extracts the information of the particle momentum p. Operators can be constructed for many different observables of a quantum system including position, energy, spin, angular momentum, etc. Dirac's foundational work on quantum mechanics, for which he won the Nobel prize, was based around his work on operators presented in the great book *Principles of Quantum Mechanics* [7]. Dirac incorporated Heisenberg's matrix mechanics, Schrödinger's wave mechanics and later his own bra-ket notation. Essentially it is possible to fully develop quantum mechanics based on a theory of operators and the relationships between them. Some physicists would go as far as to say that Dirac essentially "proved" quantum mechanics and there is nothing to worry about but simply accept it at face value and use the marvellous formalism to solve quantum problems. The quantum maxim "shut up and calculate" is well known amongst the physics community. However, some curious physicists still persist in posing "stupid" questions. They wonder fundamentally why quantum mechanics is the way it is and try to seek a deeper, conceptual understanding of quantum phenomena.

1.6.5   *Discrete quantum mechanics.*

In the famous book *Quantum Mechanics and Path Integrals* [8] the colourful Richard Feynman presented a discrete space-time derivation of the 1+1 dimension Dirac equation for a free particle. By considering just one time dimension and one space dimension the quantum problem could be considerably simplified. The model is now commonly known as the "Feynman chequerboard" – since a particle is viewed in the calculation as "zig-zagging" diagonally forwards through space-time in a similar manner to a bishop in chess. At each turn in the path the particle picks up a small phase term and by summing over all paths the Dirac propagator is found. Amazingly Feynman must have thought the derivation was too trivial since it is posed as a question to the reader who must derive it himself! Although a very clever way of discretising the Dirac equation the treatment is an approximation and assumes that the particle moves unrealistically at instantaneously the speed of light and strangely the right answer is only found in the limiting low velocity case.

However, many people have pondered over the problem ever since and it has lead to many other variants of discrete or lattice based quantum mechanics [9–13]. Notably Kauffman reworked the original chequerboard and suggested "bit string" physics to represent quantum mechanics. Also Kac and others used a discrete underlying diffusion process (again at the speed of light) to derive a continuous time differential equation, reminiscent to the 1+1 dimension Dirac equation, if mass or time is treated as an imaginary number.

Although numerous discrete approaches to quantum mechanics exist none, to my knowledge, provide an exact solution to the Dirac equation in a relativistically consistent environment. To find a discretisation approach which is exact and provides fermion features, such as spin, represents a kind of "holy grail" for quantum mechanics enthusiasts. In Chapter 2 we shall discuss such an exact model.

1.6.6 *Interpretations of quantum mechanics.*

Anybody reading this book will be familiar with the problems in interpreting the strange predictions of quantum mechanics and observed quantum phenomena such as wave-particle duality, uncertainty, spin, non-locality and more. To date all experimental evidence confirms the formalism of quantum mechanics to such an extent that the famous Aspect experiments [14], using Bell's theorem [15] to test for hidden variables, were seen by some as a bit of a waste of time since they confirmed the obvious fact – that quantum mechanics is 100% correct!

The most widely accepted interpretation these days is the Copenhagen or Bohr–Heisenberg interpretation which was developed by Niels Bohr and others in the second quarter of the 20th century. The exact Copenhagen interpretation depends on who you ask but essentially it implies that the entire quantum system is probabilistic and described by a wavefunction and no knowledge of the actual state of the system can be determined between measurements. A more exhaustive outline of the principles can be given as:

1. A quantum system is completely described by a wavefunction, which represents an observer's knowledge of the system.
2. The probability of an event is related to the square of the amplitude of the wavefunction (or the modulus squared of the wavefunction).
3. Complementarity principle – matter exhibits wave-particle duality. Experiments can measure either particle or wavelike properties of matter but not both at the same time.
4. The uncertainty principle states that is not possible to know all the values of all of the properties of the system at the same time.
5. The correspondence principle – that large quantum systems should approach classical behaviour.
6. Measuring devices are classical devices and measure classical properties of systems such as position and momentum.

Many people, although accepting the operational tenets of the Copenhagen principle, view it as somewhat vague and a "smoke screen" for new underlying physics that we do not yet understand. One could argue that virtually every proposition of the principle really demands further explanation. Other theories [16], such as involving the role of human consciousness, many worlds, pilot waves to guide particles [17] and gravitational induced collapse of the wavefunction, have all been proposed over the years. Although mostly controversial, none of these explanations has seemingly gained much traction and provided a worthy successor to the Copenhagen interpretation.

1.7 Hans Reichenbach.

Hans Reichenbach born in 1891 in Hamburg, Germany was a rare breed being both a leading philosopher and physicist. Remarkably his teachers included the famous physicists Max Planck, Max Born and Albert Einstein, the mathematician David Hilbert and the philosopher Ernst Cassirer. In 1938 he moved to UCLA where he published significant works on the axiomatisation of quantum mechanics, the philosophy of space and time and the meaning and interpretation of special relativity. Reichenbach had many interesting ideas that we will draw upon heavily in this book. Incredibly, during my six years of formal physics education at Cambridge University nobody bothered to mention him since perhaps most physicists regard Reichenbach as a bit too much of an "undesirable" philosopher.

1.7.1   *Reichenbach's interpretation of quantum physics.*

Reichenbach argued that there are only two interpretations of quantum mechanics that are free from *causal anomalies* – that is they do not violate causality under special relativity. The wave-particle duality of quantum mechanics is clearly demonstrated in the common two slit experiment and when two slits are open the movement of a photon corresponds to interference but if a single slit is monitored the photon is observed to pass thorough one slit or another. These, and similar experiments, provide causal anomalies or "action at a distance" for many interpretations of quantum mechanics. In Reichenbach's view there are only two models of quantum mechanics that are free of causal anomalies. The first is the Copenhagen interpretation which states that speaking about values of unmeasured quantities is meaningless. This dodges the issue by essentially saying we cannot ask any questions about what happens between emission from the source of the photon until it is detected and measured. Reichenbach did not particularly like this approach since it allows vague, meaningless physical statements to be made. Instead he formulated another interpretation based on three-valued logic. Instead of two truth statements, true and false, he included a third *indeterminate*. All quantum statements and the role of operators can then be formalised into logic statements and truth tables can be established for various experimental outcomes. However, in my opinion Reichenbach just replaced the vague Bohr–Heisenberg approach with a formal logic or framework with unmeasured quantities just defined as indeterminate – no extra insight into the conceptual foundations of quantum mechanics was added.

1.7.2   *Reichenbach and space and time.*

Reichenbach spent much time developing a theory of space and time based on the idea of normal causality. He demonstrated that if normal causality (no causal violations) holds then it must determine the topology of space and hence the geometry within space. This is a departure from Kantian philosophy which held that *both* a law of causality and Euclidean geometry were *a priori* truths.

Reichenbach was also careful to distinguish between two fundamental concepts of time: the order of time and the direction of flow of time [18]. He developed a definition of the time order of events based on causal relationships between them. Event A is said to occur before event B only if event A can produce a physical effect on event B. If event A acts on event B and then B acts on event C then the causal chain describes the time order of events as A before B before C (Fig. 1.5).

Figure 1.5: The time order of events A, B and C.

The direction of time is more tricky since all classical and even relativistic processes are apparently time reversible under the laws of physics. If a function $F(t)$ is a solution to a classical mechanical process, say, a collision between two billiard balls, then $F(-t)$ is also an admissible solution. This conundrum has confused everybody and various not entirely satisfactory explanations have been proposed. Reichenbach suggested that the direction of time could only be defined if we consider irreversible processes. Generally these irreversible processes are considered processes that are accompanied by an increase in entropy. A glass smashing on the floor is pretty irreversible! However, the second law of thermodynamics which describes the flow of entropy, is really based on statistical mechanics which is based on classical dynamics, which is time reversible. Reichenbach proposed that a system based on open "conjunctive" forks linking causal events provides an asymmetry which determines the direction of time based on his *principle of common cause.*

1.7.3   *Reichenbach's principle of common cause.*

The application of probability theory to causality and its relation to the direction of time was developed by Hans Reichenbach. His principle of common cause (PCC) [18] was summarized as follows: "If coincidences of the two events A and B occur more frequently than would correspond to their independent occurrence, that is, if these events satisfy $P(A.B) >$

$P(A)P(B)$ then there exists a common cause C for these events that the fork ACB is conjunctive." That is the probability of A and B occurring together is greater than the product of the individual probabilities of A and B. A conjunctive fork ACB (see Fig. 1.6) between events is open on one side where C is earlier in time than A or B. This asymmetry Reichenbach argued provides a definition of the flow of the direction of time in terms of microstatistics. Essentially a common cause is expected when coincidences or correlations between events occur repeatedly with greater frequency than complete statistical independence $P(A.B) = P(A)P(B)$. Reichenbach's original geyser example serves well to illustrate the principle: "Suppose two geysers which are not far apart spout regularly, but throw up their columns of water always at the same time. The existence of a subterranean connection of the two geysers with a common reservoir of hot water is then practically certain. The fact that the measuring instruments such as barometers always shows the same indication, if they are not too far apart, is a consequence of the existence of a common cause – here the air pressure."

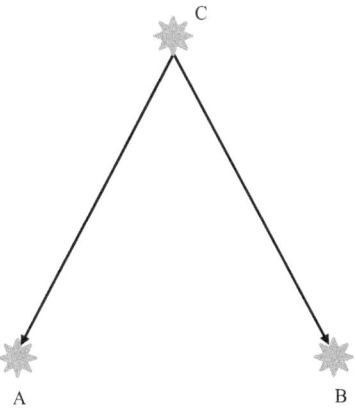

Figure 1.6: Reichenbach "conjunctive" fork linking events A and B with common cause C. C is earlier in time than simultaneous events A and B.

The principle of common cause can provide a definition of simultaneity. If A and B are simultaneous there cannot be a causal linkage between them except through the earlier event C. Reichenbach wrote on simultaneity: "The concept simultaneous is reduced to the concept indeterminate as to time order. This result supports our understanding of the concept simultaneous. Two simultaneous events are so situated that a causal chain cannot travel from one to the other in either direction ... simultaneity means the exclusion of causal connection." This is based on the relativistic principle of the maximal velocity of light c and is compatible with the concept of cause preceding effect. It is worth noting that historically this view was not always held since, for example, Descartes held that light had infinite propagation speed which was later found to be empirically false.

1.7.4   *Relativistic principle of common cause.*

Reichenbach's principle of common cause is very general and must be obviously slightly modified in a relativistic framework. The relativistic connectivity of possible events must be considered and if from Einstein's relativity the speed of light is the maximum signal speed then the light cone structure must determine the causal structure. Consider Figure 1.7 where event A can be influenced by an event at C in the past since the space-time path that connects them is "time-like" or "light-like". An event at D or E cannot interact or be causally connected with A since the path between them is "space-like" and would require a faster than light connection or signal [19].

Based on this, Penrose developed a *law of conditional independence* [20] relating the probabilities in regions of space-time that have overlapping past light cones. This is very similar to Reichenbach's PCC but considers the prior relativistic space-time regions C that have causal influence on say events A and B and provide a causal correlation between A and B.

The work of Malament is also interesting in providing a relativistic definition of simultaneity between space-time events based on the light cone structure. Malament defined a *standard simultaneity condition* [21] where actual simultaneous events lie on a hyperplane orthogonal to the particle world-line and he defined several symmetries (translation along world line, scale expansion, reflection and spatial rotations) for valid hypersurfaces of simultaneity. Interestingly these symmetries are consistent with the causal net model we shall develop in Chapter 2.

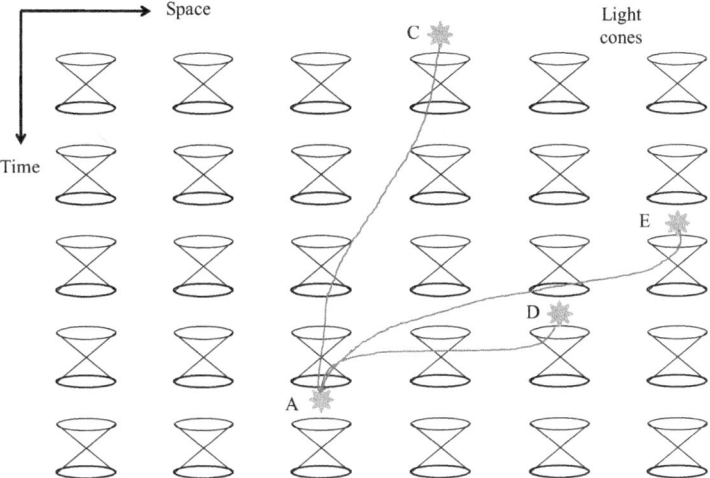

Figure 1.7: Causally linked events in Minkowski space-time.

1.8 Emergence in Physics.

At this point we will leave our discussion on causality and space-time and introduce the fashionable topic of emergence in physics. Complex structures emerge in nature from relatively simple underlying phenomena from delicate snowflakes to termite mounds and behaviour of traffic jams on the M25. Usually, these complex phenomena are totally unexpected and cannot be easily inferred from a knowledge of the underlying behaviour of the lower-level entities. Who could guess that basic water molecules could crystallise to produce beautiful snowflakes or simple termites could build huge mounds that are magnetically aligned to provide cooling in the sun? A whole cross-disciplinary field has arisen these last few years to study "emergence" in complex systems.

Figure 1.8 The extremely beautiful and complex patterns of snowflakes. © CORBIS

   The role of emergence in physics was outlined by P.W. Anderson in 1972 in his now famous essay *More is Different* [22]. He argued against the reductionist hypothesis of the "arrogant particle physicist" and pointed out that reducing everything to apparently fundamental laws does not readily imply the ability to use those laws to reconstruct the universe. The universe is full of a virtual infinity of complex processes (and beings!) of different scales that cannot be understood or predicted on the basis of the understanding of a few particles. Apparently the discover of the positron (funnily enough another Anderson – Carl Anderson Nobel laureate 1936), after his discovery, said "the rest is chemistry". However, in reality we now know that at higher and higher levels of organisation new phenomena and physical laws evolve. To illustrate Anderson proposed a hierarchy whereby elementary entities of science X obey the laws of science Y where at each stage new concepts and laws unexpectedly emerge. These different levels appear like the layers of an onion or perhaps a better analogy is a branching tree since at each level of the hierarchy many new complex phenomena and "laws" evolve.

Figure 1.9 Magnetically aligned termite mounds in Australia. Very clever termites!
© CORBIS

Table 1.2: Anderson's hierarchy of the laws of science.

X	Y
solid state or many-body physics	elementary particle physics
chemistry	many-body physics
molecular biology	chemistry
cell biology	molecular biology
•	•
•	•
•	•
psychology	physiology
social sciences	psychology

Over the years, to the dislike of many physicists, even the so-called "fundamental" laws of physics have not been spared. Following the discovery of Newton's laws the idea of a "clock-work" universe, following entirely predictable deterministic laws, held sway with God viewed as the precision "clock-maker" himself. The discovery of quantum mechanics, with its inherent indeterminism, demonstrated that Newton's laws, although beautiful, were but emergent. Microscopic matter, such as atoms and molecules, does not behave like Newtonian billiard balls but exhibits new unexpected phenomena such as diffraction. Classical physics arises from the aggregation of microscopic matter into macroscopic solids, fluids and gases. A correspondence principle was developed whereby Newton's laws emerge in the macroscopic limit, albeit in a slightly vague way (Section 1.6.6).

Thus, at higher and higher levels of organisation new phenomena and physical laws evolve. The quantum laws of electron motion provide the Newtonian based laws of statistical mechanics providing the laws of thermodynamics and chemistry. These laws lead to the laws of crystallisation and laws of materials (rigidity, plasticity, ...) and eventually the laws of building, construction and civil engineering which even dictate town planning! The laws at each level are remarkably self contained and hold over an incredibly wide range of underlying variables. A chemist does not need to know about quarks and the Schrödinger equation to analyse chemical reactions that can be remarkably analysed in a symbolic form, for example $CH_4 + 2\,O_2 \rightarrow CO_2 + 2\,H_2O$ (the burning of methane) or the "skeletons" and "curly arrows" of advanced organic chemistry. An engineer building a bridge uses engineering equations to calculate the load on concrete and the tensile properties of steel beams and does not need to involve himself in chemical reactions. Fortunately for students you can choose to study a discipline such as physics, chemistry, geology, engineering or computer science without worrying about the other subjects! It is amazing how compartmentalised yet interdependent these laws are. A particle physicist might think that he "knows the laws of the universe and everything in it" as Sheldon in the *Big Bang Theory* but in reality they only have a ground floor view ... and there might even be a basement(s) we don't know about.

1.8.1   *Emergent quasiparticles in nature.*

A particularly flourishing field for emergent phenomena for the past 30 years is solid state or condensed matter physics. For the Nobel Prize winner Laughlin, in his great book *A Different Universe* [23], the case for emergence became overwhelming when von Klitzing discovered the quantum Hall effect. In the quantum Hall effect the Hall resistance has steps that correspond to universal quantised values. That macroscopic matter could display quantised behaviour in units related to the quantum Planck's constant is quite amazing. A similar phenomenon is the macroscopic coherent quantum behaviour demonstrated by superconductors. Laughlin himself discovered the extraordinary fractal quantum Hall effect where elementary excitations, now called Laughlin quasiparticles, with precise fractions, such as $e/3$, of the supposedly elementary electron charge e were observed. Numerous so called quasiparticles, such as phonons, holes, magnons, etc. have been observed in condensed matter physics (see Table 1.3). These are elementary excitations of a substance or media that behave almost perfectly like individual particles. For example, sound waves are transmitted in solids by lattice vibrations or quasiparticle phonons and it is even possible to detect the movement of a single discrete phonon. The physics of quasiparticle holes in semiconductors is fundamental to the development of the microprocessor and the modern computer. More recently, magnetic monopoles have been observed at low temperatures in special "frustrated" magnetic materials called spin-ice (actually pyrochlore lattices) [24]. These monopoles are not fundamental but quasiparticles that emerge from the correlation properties of the surrounding medium. Incredibly the physical equations they describe are directly analogous to that of the electron where the magnetic charge replaces the electric charge – they even apparently obey a form of Ohm's law!

Table 1.3: A few emergent quasiparticles in condensed matter physics.

Quasiparticle	Description
"electron"	Electrons in solids can exhibit different effective mass and charge from in a vacuum.
hole	An empty electron state in a semiconductor valence band with positive charge.
phonon	Quantum of a sound wave. Collective excitation or vibration of atoms in a lattice.
magnon	Quantum of a spin wave in a magnetic material.
plasmon	Quantum of plasma oscillations. Collective excitation in a plasma or of ions.
polaron	Quasiparticle from interaction of electron with surrounding ions.
magnetic monopole	Spin "flips" that propagate in frustrated magnetic systems.

Current thinking is that the 12 discovered fermions, such as the electron and other leptons and the quarks, are indivisible and truly fundamental. Experimentally there appears to be no internal structure to them. Are they really fundamental or are they emergent quasiparticles? They are described by the Dirac equation and does this represent a fundamental equation of physics? Is it the "basement" equation or just an approximation, as the Schrödinger equation, valid within certain physical limits? Perhaps even quantum mechanics itself is an emergent phenomenon of something simpler and more fundamental? The conceptual problems of wave-particle duality and quantum measurement perhaps disappear when we look at a different layer of the emergent "onion"? In the next chapter we shall explore these concepts and provide a theory possibly more fundamental than quantum mechanics and special relativity – a theory of space-time constructed entirely from the concept of causality and probability.

1.8.2   *Emergent universes.*

Entrepreneur and physicist Stephen Wolfram, in his heavy-weight tome *A New Kind of Science* [25], took the emergence idea to its limit by arguing that the universe is an emergent phenomenon. Based on his empirical work on cellular automata computer programs he suggested that the universe and everything it contains is digital and could perhaps emerge from very simple computer algorithms. The laws of physics would thus develop from a simple starting program that would evolve into the complex hierarchy of laws. Obviously Wolfram has no idea what this program might be but extremely rich and unexpected behaviour seems to be generated from remarkably simple cellular automata.

The idea of emergent universes is perhaps not so far-fetched that serious philosophers, such as Nick Bostrom, actually publish on them. Although it may sound like the *Matrix* movie, the famous *simulation argument* [26] proposed that for a massively technologically advanced alien civilisation, possessing almost infinite cheap computing power, the creation of artificial universes would be trivial. If this were the case then it is more probable that we live in a simulated universe rather that the "real" universe. In my view this is in the realm of science fiction but when considering emergence one possible logical conclusion is that the whole of nature and hence the universe is emergent from "something" truly fundamental and in any case the laws of physics emerge from a very simple process.

Chapter 2

A Theory of Causal Space-Time

2.1 Introduction

In Chapter 2 of this book we shall attempt to construct a theory of physics based on the fundamental concepts of causality and probability. Regrettably, the treatment is rather mathematical but I have attempted to present it in the simplest form possible without too much formalism and conveying the main concepts involved. Indeed for certain people the treatment is perhaps not mathematically rigorous enough. If necessary the reader can skip to Chapter 3 for a discussion of the main results.

2.2 Space-Time Causal Nets.

We shall adopt a relational view of time, as Leibniz, as an ordered series of closely spaced "events". What constitutes an event is a hotly discussed issue amongst philosophers but for our purposes represents an indivisible point or "something" in space-time – perhaps a single "bit" or "pixel" of information. Physicists often use the notion of events when talking about special relativity without worrying too much about the philosophy of what they are – just the minimum entity that can "exist", say, a flash of light or a photon traversing a point of space-time.

Now if we consider time as a series of closely spaced events then from this perspective a classical particle trajectory could appear as a statistically correlated series of events in space-time (for example, a series of actual observations). If the correlation is perfect then one may loosely say that an event at one point in space "causes" the event at the next point, providing a Newtonian trajectory (Fig. 2.1). For example, if a Newtonian particle moves through space-time we can predict or know where the particle will be from one moment to the next. Experience tells us that we can easily predict the trajectory of a thrown cricket ball from one moment to the next with a knowledge of its previous motion and any forces acting on it.

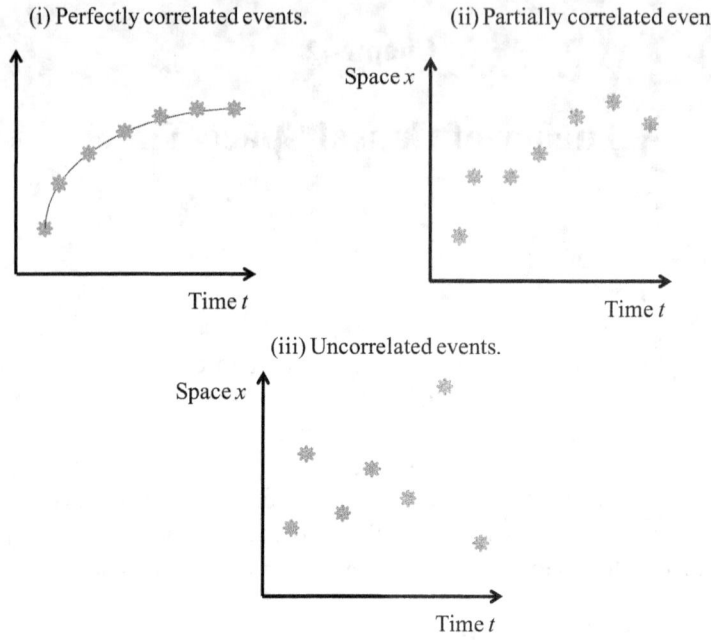

Figure 2.1: Correlated and uncorrelated events in space-time.

However, if the correlation of events is imperfect, but greater than that resulting from statistical independence, then adjacent events in space are implied to have a common cause originating at a previous time [9]. Consider a particle that from one instant to the next has a randomness or uncertainty in its position but is obviously related to or correlated to its position at an earlier time. A cricket ball with this type of motion would be slightly unpredictable in its movement and we would not be able to precisely determine its trajectory. The trajectory becomes probabilistic in nature and we would have to involve a statistical interpretation. We might suppose that the event at an earlier time could be considered to provide a common cause analogous to the common cause discussed by Reichenbach (Section 1.7.3).

Reichenbach argued in quite general terms that a satisfactory definition of time could only be obtained on the basis of a principle of common cause [18]. By this he meant that if two events A and B occur repeatedly with greater frequency than complete statistical independence predicts (i.e. $P(A.B) > P(A)P(B)$) then there exists an event C at previous time such that the fork ACB is conjunctive, or has one side open (see Fig. 1.6). A network of such conjunctive forks constitutes a causal network in which time is ordered and events may be considered simultaneous only when they share a common cause.

In discussing the correlation between events we have introduced the notion of probability. We can apply the conventional Laplacian definition of probabilities based on the observed frequency of events divided by the possible number of events. Jaynes [27] demonstrated that probability theory, the mathematics of probabilities, could be essentially derived from logic and common sense. In any case, to further develop our causal net we need to apply probability theory.

To construct the causal net for a particle motion in space-time, we consider a 1-dimensional space aligned with the direction of particle motion, and embedded in 3-dimensional space. In this 1-dimensional space the simplest causal net that satisfies our definition of simultaneity is a 1+1 dimensional "diamond" lattice with causal links connecting the lattice points as in Figure 2.2.

Each causal connection is defined by a connecting arrow giving a definite lineal order and an associated probability. Each vertex on the causal net represents a possible event – meaning a possible observation of the particle – and has two incoming and two outgoing causal connections so that each event has two effective possible common causes. Starting at a vertex and following an outgoing arrow at random at each subsequent vertex describes a "causal chain" as a series of possible events. The causal net thus describes the connectivity between all causal connections and can statistically model all future possibilities.

Measurement or observation at a vertex or a region of the net provides, through Bayesian statistics, a re-evaluation of these probabilities after a measurement. For example, if we possess no knowledge of where an actual event might occur on the net but observe an event at a particular location then, due to the connectivity of the net, we can say that certain causal chains or paths could have lead to the event and that it was more likely that the event was preceded by an event in a cone or region of previous possible events. This is illustrated in Figure 2.2 where an event at A is more likely to have been caused by an event at B than C and D is an impossibility due to zero connectivity between the paths. Bayes theorem provides a way of translating this common sense concept into a formal probabilistic context since $P(A|B) > P(A|C) > P(A|D)$.

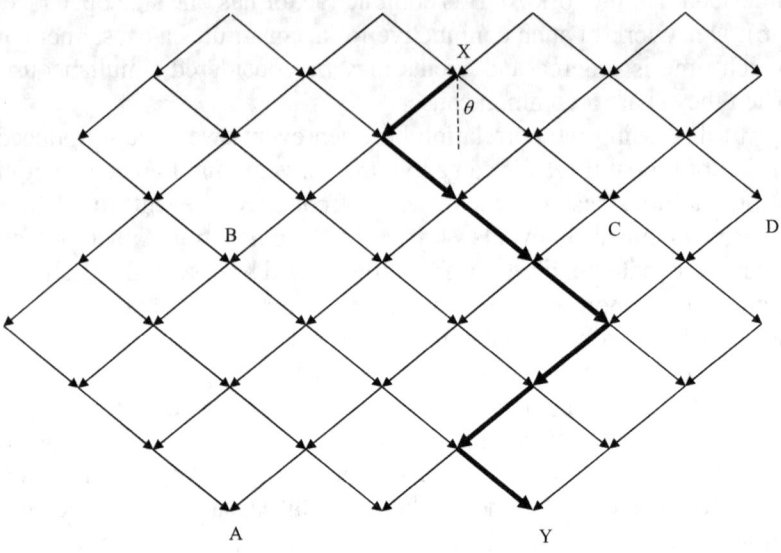

Figure 2.2: Causal net showing causal chain from X to Y.

2.3 Relativistic Causal Nets for a Free Particle.

First we will consider the simple case of a particle randomly diffusing on the causal net shown in Figure 2.2. In this model time and space can be discretely "counted" by attaching an integer to each of the vertex points but there is no underlying continuous space-time. To relate to conventional mechanics we interpolate this set of integers by a set of real number coordinates. Expecting that space and time have different dimensions we need to introduce a constant c with dimensions [space/time]. The net is then made up of elementary triangles labelled with $(\Delta x, c\Delta t, c\Delta \tau)$ as shown in Figure 2.3. We have not yet added any specific interpretation to these quantities. To guarantee invariance of causality on the net we impose c as the speed of light [28]. Since, from geometry, $\frac{\Delta x}{c\Delta t} = sin\theta \leq 1$, we then identify Δx and Δt as relativistic space-time intervals in an observer frame S and $\Delta\tau$ as the particle proper time interval in its rest frame S'. The net geometry guarantees the invariant space-time interval

$$(c\Delta\tau)^2 = (c\Delta t)^2 - (\Delta x)^2.$$

(2.1)

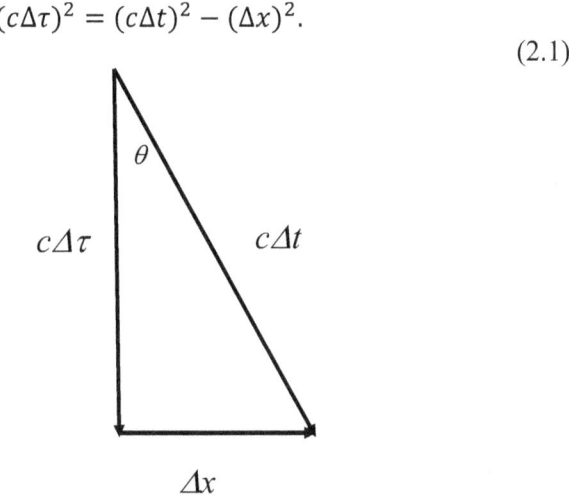

Figure 2.3: The elementary space-time "triangle" for the causal net.

Having abandoned the concept of absolute and continuous space-time we need to define the observed velocity in terms of finite differences. The definition we shall adopt is $v = \Delta x/\Delta t$ which we equate to the expectation of the velocity on the causal net. The two time intervals are then related by $\Delta\tau = \Delta t/\gamma$ where γ is the Lorentz factor

$$\gamma = 1/\sqrt{1 - v^2/c^2}$$

specifying the net angles

$$\cos\theta = \frac{1}{\gamma}, \qquad \sin\theta = \frac{v}{c}$$

We now specialise to the case of the motion of a free particle. Clearly Eq. (2.1) and thus the net can be scaled by a factor. If we identify this with the particle rest mass m then rearranging Eq. (2.1) we then have the relativistic dispersion relation $E^2 = p^2c^2 + m^2c^4$ where E is the particle energy $E = \gamma mc^2$ and $p = \gamma mv$ the momentum.

By construction we require the lattice to describe only physically admissible motions of the particle. Experience shows that real particle trajectories obey a principle of least action – that is the integral $\int p\,dx$ is stationary. This is a restatement of Maupertuis principle which is a weak form of the well known principle of least action and is experimentally verified in classical mechanics. On our net if the action $\sum p\Delta x$ differed for different trajectories then this would rule some trajectories as physically inadmissible. Therefore we conclude that $\sum p\Delta x$ is the same on the lattice for all paths between two points which means that $p\Delta x$ is a constant η for a valid causal net.

2.4 Free Particle Motion on the Causal Net.

To impose our imperfect correlation of events we shall assume that there is an indeterminism or randomness to the particle motion at each net vertex. We shall make the assumption that this indeterminism is governed by Eq. (2.1) on the causal net. Thus, a particle in its own rest frame S' over interval $\Delta\tau$ moving at a speed $|v|$ in frame S can move to a position $\pm\Delta x$ in time Δt. This produces a random trajectory in space-time (see Fig. 2.2).

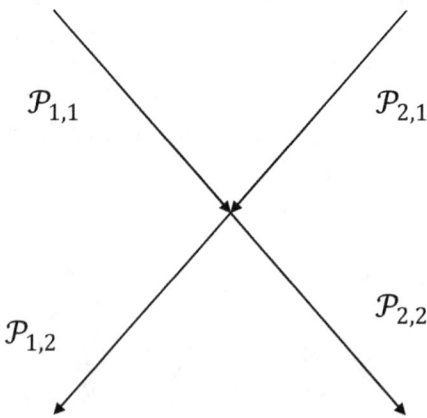

Figure 2.4: A vertex (1,2) on the causal net with associated probabilities.

Initially, we shall consider "classical" or non quantum probabilities in construction of the causal net. Consider an individual vertex on the net and label the incoming probabilities on row 1 $\mathcal{P}_{1,1}$ and $\mathcal{P}_{2,1}$ and outgoing probabilities on row 2 $\mathcal{P}_{1,2}$ and $\mathcal{P}_{2,2}$ (Fig. 2.4). Probability is conserved at the vertex and the total probability at a vertex is given by $\Omega_{1,2} = \mathcal{P}_{1,1} + \mathcal{P}_{2,1}$. If the average velocity measured on the lattice is uniform then $\mathcal{P}_{1,1} = \mathcal{P}_{2,2}$ and $\mathcal{P}_{1,2} = \mathcal{P}_{2,1}$. This implies that the probabilities "cross" at each vertex without actually interfering although the probabilities are coupled. We shall see that this corresponds to the equilibrium case of a free particle. If we consider normalised branching probabilities at the vertex defined as $\hat{\mathcal{P}}_{1,1} + \hat{\mathcal{P}}_{2,1} = 1$ then since expected velocity at the vertex is defined to be v we have

$$\mathbb{E}[v] = \gamma \frac{\Delta x}{\Delta t} \left[\hat{\mathcal{P}}_{1,1} - \hat{\mathcal{P}}_{2,1} \right] = v.$$

(2.2)

The branching probabilities are then given by

$$\hat{\mathcal{P}}_{1,1} = \frac{E + mc^2}{2E}, \qquad \hat{\mathcal{P}}_{2,1} = \frac{E - mc^2}{2E}.$$

(2.3)

From this we can see that in the low velocity limit $|v| \to 0$ then $\hat{\mathcal{P}}_{1,1} \to 1$ and $\hat{\mathcal{P}}_{2,1} \to 0$ and in the high velocity limit $|v| \to c$ then $\hat{\mathcal{P}}_{1,1} \to \hat{\mathcal{P}}_{2,1} \to 1/2$. The branching ratio Γ can be written as a function of γ or the net angle θ.

$$\Gamma = \frac{\hat{\mathcal{P}}_{1,1}}{\hat{\mathcal{P}}_{2,1}} = \frac{E + mc^2}{E - mc^2} = \frac{\gamma + 1}{\gamma - 1} = \frac{1 + \cos\theta}{1 - \cos\theta}$$

(2.4)

Using the branching probability (Eq. 2.4) we can write a non trivial matrix equation linking the probabilities

$$\mathcal{P} = \begin{pmatrix} \mathcal{P}_{1,1} \\ \mathcal{P}_{2,1} \end{pmatrix} = \begin{pmatrix} 0 & \Gamma \\ 1/\Gamma & 0 \end{pmatrix} \begin{pmatrix} \mathcal{P}_{1,1} \\ \mathcal{P}_{2,1} \end{pmatrix}.$$

(2.5)

2.5 Relativistic Quantum Mechanics on the Causal Net.

We shall now see how the causal net is compatible with the quantum mechanics of the Dirac equation for a free particle. We notice that identifying the net constant η with Planck's constant h provides the de Broglie relation $\lambda p = h$ [4] with $\Delta x = \lambda/2$, and a Heisenberg like relation $\Delta p \Delta x \sim h/2$ [29,30]. The discrete nature of the net automatically entails a de Broglie relation and an uncertainty principle.

Now a very general way of forming the probabilities $\mathcal{P}_{i,j}$ for the branch (i,j) is through a vector dot product $\mathcal{P}_{i,j} = \boldsymbol{\phi}_{i,j} \cdot \boldsymbol{\phi}_{i,j}$ with $\boldsymbol{\phi}_{i,j} = \sqrt{\mathcal{P}_{i,j}} \, (a(\tau), b(\tau))$ with real components that depend on the proper time τ at the net vertices. The probability is invariant in the rest frame of the particle and equivalent to a gauge relationship that conserves probability in S' so using the relativistic invariance $-mc^2 \Delta \tau = p\Delta x - E\Delta t$ we can write the vector as

$$\boldsymbol{\phi}_{i,j} = \sqrt{\mathcal{P}_{i,j}} \begin{pmatrix} a(\tau) \\ b(\tau) \end{pmatrix} = \sqrt{\mathcal{P}_{i,j}} \begin{pmatrix} \cos(mc^2\tau/\hbar) \\ \sin(mc^2\tau/\hbar) \end{pmatrix}$$

$$= \sqrt{\mathcal{P}_{i,j}} \begin{pmatrix} \cos\left((px - Et)/\hbar\right) \\ \sin\left((px - Et)/\hbar\right) \end{pmatrix}.$$

However, a more conventional and compact way of forming the probabilities is through introducing complex numbers, rather than vectors, to carry the phase information. So instead by combining complex probability amplitudes we can write $\mathcal{P}_{i,j} = \phi_{i,j} \cdot \phi_{i,j}^*$ with

$$\phi_{i,j} = \sqrt{\mathcal{P}_{i,j}} \, e^{-\frac{imc^2\tau}{\hbar}},$$

$$(2.6)$$

which depends on the proper time τ at the net vertices. Notably the phase is independent of position x for a particular τ. Again using the relativistic invariance we can write the probability amplitude as

$$\phi_{i,j} = \sqrt{\mathcal{P}_{i,j}} \, e^{-\frac{imc^2\tau}{\hbar}} = \sqrt{\mathcal{P}_{i,j}} e^{\frac{i(px-Et)}{\hbar}},$$

$$(2.7)$$

Where x and t are defined at the discrete net vertices and for the moment we consider only positive roots. We can then rewrite Eq. (2.5) as

$$\boldsymbol{\phi} = \begin{pmatrix} \phi_{1,1} \\ \phi_{2,1} \end{pmatrix} = \begin{pmatrix} 0 & \sqrt{\Gamma} \\ 1/\sqrt{\Gamma} & 0 \end{pmatrix} \begin{pmatrix} \phi_{1,1} \\ \phi_{2,1} \end{pmatrix},$$

(2.8)

which can be alternatively expressed for Eq. (2.7) in terms of a *unique* transfer matrix $\boldsymbol{M}$

$$\boldsymbol{\phi} = \begin{pmatrix} \phi_{1,1} \\ \phi_{2,1} \end{pmatrix} = \boldsymbol{M} \begin{pmatrix} \phi_{1,1} \\ \phi_{2,1} \end{pmatrix},$$

(2.9)

defined as

$$\boldsymbol{M} = \begin{pmatrix} \cos\theta & \sin\theta \\ \sin\theta & -\cos\theta \end{pmatrix} = \begin{pmatrix} 1/\gamma & v/c \\ v/c & -1/\gamma \end{pmatrix} = \frac{1}{E} \begin{pmatrix} mc^2 & pc \\ pc & -mc^2 \end{pmatrix} = \frac{H_D}{E}.$$

(2.10)

Here we recognise H_D as the Dirac Hamiltonian for a free particle [3, 21] with defined momentum p. To connect with the complete quantum mechanics we note that Eq. (2.10) can be put in the conventional form [5] by assuming that space-time is locally differentiable at the vertex, allowing us to use the usual momentum operator $\hat{p}$ to replace the momentum eigenvalues p writing

$$\begin{pmatrix} mc^2 & c\hat{p} \\ c\hat{p} & -mc^2 \end{pmatrix} \boldsymbol{\Psi} = E\boldsymbol{\Psi} = i\hbar \frac{\partial \boldsymbol{\Psi}}{\partial t},$$

(2.11)

where we have replaced the probability amplitudes $\boldsymbol{\phi}$ with the familiar 2 component Dirac spinor $\boldsymbol{\Psi}$ for the free particle [31]

$$\boldsymbol{\Psi}(x,t) = \begin{pmatrix} \psi_1 \\ \psi_2 \end{pmatrix} = \boldsymbol{\phi} = \begin{pmatrix} \phi_{1,1} \\ \phi_{2,1} \end{pmatrix} = A \begin{pmatrix} 1 \\ \dfrac{pc}{E+mc^2} \end{pmatrix} e^{\frac{i(px-Et)}{\hbar}}$$

(2.12)

and A is an appropriate normalisation constant.

2.6 Causal Net Quantum Symmetries and Spin.

Note that the unique matrix $\boldsymbol{M}$ above is a unitary, orthogonal matrix which provides an SU(2) group transformation corresponding to an improper rotation – that is a rotation $\boldsymbol{R}(\theta)$ followed by an inversion $\boldsymbol{\beta}$ so $\boldsymbol{M} = \boldsymbol{\beta R}(\theta)$. The matrix provides the transformations for the probabilities $\mathcal{P} = \boldsymbol{M}^2\mathcal{P} = \boldsymbol{I}\mathcal{P}$ and probability amplitudes $\boldsymbol{\Psi} = \boldsymbol{M\Psi}$. Importantly because $\boldsymbol{M}^3 = \boldsymbol{M}$ there automatically exists only two levels of symmetry at the vertex and the causal net provides simultaneously both the probabilities and the underlying probability amplitudes. Since it is an improper rotation the symmetry determines a preferred axis which provides helicity along the axis of movement. If we revisit Eq. (2.7) and consider both possible positive and negative roots we can see that even and odd solutions that provide helicity $\lambda = \pm 1$ with positive energy $\varepsilon = +1$ are given by

$$\boldsymbol{\phi}_{\substack{\varepsilon=+1 \\ \lambda=+1}} = \begin{pmatrix} \sqrt{\mathcal{P}_{1,1}} \\ \sqrt{\mathcal{P}_{2,1}} \end{pmatrix} \qquad \boldsymbol{\phi}_{\substack{\varepsilon=+1 \\ \lambda=-1}} = \begin{pmatrix} \sqrt{\mathcal{P}_{1,1}} \\ -\sqrt{\mathcal{P}_{2,1}} \end{pmatrix}$$

$$(2.13)$$

corresponding to transfer matrices $\boldsymbol{M}_{\varepsilon=+1,\lambda=\pm1} = \boldsymbol{\beta R}(\pm\theta)$. Note that in the above and following discussion we have omitted the phase factor $e^{-imc^2\tau/\hbar}$ since this cancels in both sides of Eq. (2.9).

2.7 Negative Energy States.

Until now we have considered only the positive energy states, but negative energy solutions arise from the negative solution of the relativistic dispersion relation $E = \varepsilon\sqrt{p^2c^2 + m^2c^4} = \varepsilon|E| = \varepsilon\gamma mc^2$ ($\varepsilon = \pm 1$). This results in a reversal of the branching probabilities in Eq. (2.3) and two additional possible even and odd spinor solutions

$$\boldsymbol{\phi}_{\substack{\varepsilon=-1 \\ \lambda=+1}} = \begin{pmatrix} -\sqrt{\mathcal{P}_{2,1}} \\ \sqrt{\mathcal{P}_{1,1}} \end{pmatrix} \qquad \boldsymbol{\phi}_{\substack{\varepsilon=-1 \\ \lambda=-1}} = \begin{pmatrix} \sqrt{\mathcal{P}_{2,1}} \\ \sqrt{\mathcal{P}_{1,1}} \end{pmatrix}$$

$$(2.14)$$

for transfer matrices $M_{\varepsilon=-1,\lambda=\pm1} = -\beta R(\pm\theta)$. These states provide inverted branching ratios in Eq. (2.4) so in our model negative energy states correspond to particles moving in the opposite spatial direction or with negative velocity. This implies that either the negative energy solutions are inappropriate in our framework or the net should be perhaps redefined as speed rather than velocity with $\mathbb{E}[v] = \pm|v|$ instead of Eq. (2.2). These negative energy solutions however, have a phase that evolves in the opposite sense with proper time in Eq. (2.6).

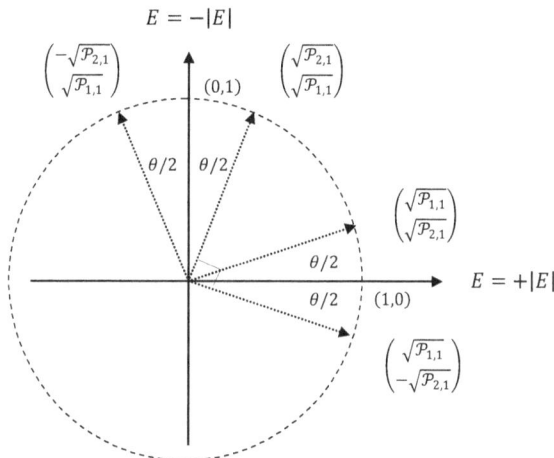

Figure 2.5: Representation of 1+1 dimension causal net solutions (Eq. 2.13 and Eq. 2.14) to the Dirac equation.

2.8 The Foldy–Wouthuysen Transformation.

The causal net is also consistent with the extraordinarily simple Foldy–Wouthuysen representation [32] of the Dirac equation where the positive and negative energy states are decoupled through a rotation of θ (the lattice angle) $R(\theta)$ of the Dirac Hamiltonian. For example, one Foldy–Wouthuysen state is given by the rotation through $\theta/2$ of the Dirac state Eq. (2.12)

$$\phi_{\substack{FW \\ \varepsilon=+1}} = R(\theta/2)\phi_{\substack{Dirac \\ \varepsilon=+1 \\ \lambda=+1}} = \begin{pmatrix} \sqrt{\mathcal{P}_{1,1}} & \sqrt{\mathcal{P}_{2,1}} \\ -\sqrt{\mathcal{P}_{2,1}} & \sqrt{\mathcal{P}_{1,1}} \end{pmatrix} \begin{pmatrix} \sqrt{\mathcal{P}_{1,1}} \\ \sqrt{\mathcal{P}_{2,1}} \end{pmatrix} = \begin{pmatrix} 1 \\ 0 \end{pmatrix}.$$

$$(2.15)$$

This rotated state and the previous Dirac states (Eq. 2.12 and Eq. 2.13) can be represented more clearly as vectors in the Figure 2.5. The Foldy–Wouthuysen states correspond with our framework where the Hamiltonian and velocity operator satisfy their classical analogues. Importantly, in this representation, establishing an exact particle position is impossible (there is only a mean position operator) and a particle is viewed as spread out over a finite region of about a wavelength which is consistent with our causal net picture since we cannot localise a particle between two adjacent net vertices without an averaging over vertices being performed in a measurement.

2.9 The Complete 4-Vector Dirac Equation in 1+1 Dimensions.

If we include the negative energy states then, by combining all 4 net solutions above (Eq. 2.13 and Eq. 2.14), we can write 4 orthogonal 4-vectors which for helicity $\lambda = \pm 1$ and $\varepsilon = \pm 1$ are

$$\Psi_{p,\varepsilon,\lambda=+1}(x,t) = A\sqrt{\frac{E+mc^2}{2E}} \begin{pmatrix} 1 \\ 0 \\ \dfrac{cp}{E+mc^2} \\ 0 \end{pmatrix} e^{\frac{i(px-Et)}{\hbar}}$$

$$E = \varepsilon|E|, \lambda = +1$$

$$\Psi_{p,\varepsilon,\lambda=-1}(x,t) = A\sqrt{\frac{E+mc^2}{2E}} \begin{pmatrix} 0 \\ 1 \\ 0 \\ \dfrac{-cp}{E+mc^2} \end{pmatrix} e^{\frac{i(px-Et)}{\hbar}}$$

$$E = \varepsilon|E|, \lambda = -1$$

$$(2.16)$$

where A is again an appropriate normalization constant. Using the 4 possible transfer matrices $M_{\varepsilon=\pm1,\lambda=\pm1}$ then the 1+1 dimension 4-matrix Dirac equation Eq. (2.15) is

$$\begin{pmatrix} mc^2 & c\beta\hat{p} \\ c\beta.\hat{p} & -mc^2 \end{pmatrix} \Psi = E\Psi = i\hbar\frac{\partial\Psi}{\partial t}.$$

(2.17)

Now this Dirac equation and the spinor wavefunction Eq. (2.16) correspond to exactly the conventional 3+1 dimension Dirac spinor for the special case of the particle moving along the x-axis and with a well-defined spin (helicity) aligned parallel and antiparallel with the x-axis [5,6].

2.10 The 3+1 Dimensional Dirac Equation.

To extend to the general 3+1 dimension case we must consider transformations of the causal net that leave it invariant under variation of direction of velocity **v** of the frame S. Using polar coordinates then for momentum $\boldsymbol{p} = |\boldsymbol{p}|(sin\vartheta cos\varphi, sin\vartheta sin\varphi, cos\vartheta)$ and we can expect that the wavefunction components become dependent on the coordinates (ϑ, φ) so $\sqrt{\mathcal{P}_{i,j}}$ becomes $\sqrt{\mathcal{P}_{i,j}}\chi(\vartheta, \varphi)$. Using a reduced vector notation $\left(cos\vartheta, sin\vartheta e^{i\varphi}\right)$ and following Dirac's convention [6] we can replace the 1-dimensional momentum operator $\hat{p}$ with the 3-dimensional momentum operator in Eq. (2.17)

$$\hat{\sigma}.\hat{p} = \begin{pmatrix} p_z & p_x - ip_y \\ p_x + ip_y & -p_z \end{pmatrix} = |\boldsymbol{p}|\begin{pmatrix} cos\vartheta & sin\vartheta e^{-i\varphi} \\ sin\vartheta e^{i\varphi} & -cos\vartheta \end{pmatrix}$$

(2.18)

formed from Pauli matrices σ_k

$$\sigma_1 = \begin{pmatrix} 0 & 1 \\ 1 & 0 \end{pmatrix}, \quad \sigma_2 = \begin{pmatrix} 0 & -i \\ i & 0 \end{pmatrix}, \quad \sigma_3 = \begin{pmatrix} 1 & 0 \\ 0 & -1 \end{pmatrix}.$$

(2.19)

Eq. (2.18) is an improper rotation and by definition the provides the relation $(\hat{\sigma}.\hat{p})\chi_{\pm} = \pm|p|\chi_{\pm}$ with two eigenvectors

$$\chi_{+} = \begin{pmatrix} \cos\frac{\vartheta}{2} \\ \sin\frac{\vartheta}{2}e^{i\varphi} \end{pmatrix} \quad , \quad \chi_{-} = \begin{pmatrix} -\sin\frac{\vartheta}{2}e^{-i\varphi} \\ \cos\frac{\vartheta}{2} \end{pmatrix}$$

(2.20)

The general solutions for the wavefunction then become (from Eq. 2.12 and Eq. 2.13) four 4-component orthogonal vectors corresponding to up and down spin $S = \pm1/2$ with positive and negative energy $\varepsilon = \pm1$. Omitting the phase factors $e^{-imc^2\tau/\hbar}$ and normalisation constant these are

$$\Psi_{\substack{\varepsilon=+1 \\ S=+1/2}} = \begin{pmatrix} \sqrt{\mathcal{P}_{1,1}}\chi_{+} \\ \sqrt{\mathcal{P}_{2,1}}\chi_{+} \end{pmatrix} \qquad \Psi_{\substack{\varepsilon=+1 \\ S=-1/2}} = \begin{pmatrix} \sqrt{\mathcal{P}_{1,1}}\chi_{-} \\ -\sqrt{\mathcal{P}_{2,1}}\chi_{-} \end{pmatrix}$$

$$\Psi_{\substack{\varepsilon=-1 \\ S=+1/2}} = \begin{pmatrix} -\sqrt{\mathcal{P}_{2,1}}\chi_{+} \\ \sqrt{\mathcal{P}_{1,1}}\chi_{+} \end{pmatrix} \qquad \Psi_{\substack{\varepsilon=-1 \\ S=-1/2}} = \begin{pmatrix} \sqrt{\mathcal{P}_{2,1}}\chi_{-} \\ \sqrt{\mathcal{P}_{1,1}}\chi_{-} \end{pmatrix}$$

(2.21)

These are the general solutions to the conventional 3+1 dimension Dirac equation

$$\begin{pmatrix} mc^2 & c\hat{\sigma}.\hat{p} \\ c\hat{\sigma}.\hat{p} & -mc^2 \end{pmatrix}\Psi = E\Psi = i\hbar\frac{\partial\Psi}{\partial t},$$

(2.22)

which can also be written in the well known [5] form

$$[c\hat{\alpha}.\hat{p} + mc^2\hat{\beta}]\Psi = E\Psi = i\hbar\frac{\partial\Psi}{\partial t},$$

(2.23)

using the matrices $\hat{\alpha}$ and $\hat{\beta}$ defined as

$$\alpha_k = \begin{pmatrix} 0 & \sigma_k \\ \sigma_k & 0 \end{pmatrix}, \qquad \hat{\beta} = \begin{pmatrix} I & 0 \\ 0 & -I \end{pmatrix}.$$

(2.24)

Thus for a 3 dimensional space we require all 3 Pauli matrices to construct the vector $\hat{\sigma}$ and the dimensionality of space defines the Pauli matrices. If we rotate Eq. (2.22) by $(-\vartheta/2)$ then we see that the net is constructed along the momentum direction $\left(\cos\vartheta, \sin\vartheta e^{i\varphi}\right)$ and that the causal net transfer matrix in Eq. (2.10) is invariant. Velocity or momentum of the particle at any point in space-time only has one direction even though it is embedded in 3-dimensional space and a plane wave solution to the Dirac equation has a unique velocity direction. The connectivity of the plane wave causal net itself only requires one space axis. The specification of the Pauli scheme is not unique and we could use a different matrix system or permutation and combining with Eq. (2.9) this allows 12 different permutations.

2.11 Effect of Potential on Causal Net and the Pauli Equation.

The case we have examined is that of a free particle but we could include a potential V on the causal net since E can be replaced by $E-V$ in the construction of the lattice and the branching ratios. Returning to the 1+1 dimension case, between two media with different potentials the net is compressed or stretched in space in the potential region with a form similar to Snell's law $\cos\theta_2/\cos\theta_1 = E/(E-V)$. We can write Eq. (2.10) conveniently as

$$M = \begin{pmatrix} \cos\theta_2 & \sin\theta_2 \\ \sin\theta_2 & -\cos\theta_2 \end{pmatrix} = \frac{1}{E-V}\begin{pmatrix} mc^2 & pc \\ pc & -mc^2 \end{pmatrix} = \frac{H_D}{E-V}.$$

(2.25)

Recall that an incremental change in potential is equivalent to a force acting on the particle so our net can encompass the full mechanics of the particle. The probability current is conserved at a potential barrier if we consider the relativistic change in probability across the barrier arising from a Lorentz contraction/expansion.

Taking these ideas a bit further we can consider that introducing a phase in Eq. (2.6) is equivalent to a global gauge invariance represented by a unitary group U(1). The Lagrangian for our net can be written in the form $\phi^+ M\phi - \phi^+\phi = 0$. What happens if we consider invariance under a local gauge transformation? For example, if we consider an SU(2) matrix, such as a rotation $U = R(\zeta)$, then in general $M(U\phi) \neq (U\phi)$ so to retain invariance we must add an additional term to the Lagrangian and the Dirac equation which corresponds to an electromagnetic potential term. We can see that a small change ζ in net angle relates to a vector potential providing an apparent force. To illustrate, consider the special case of a transformation where the proper time interval $\Delta\tau$ is unchanged by a potential. The triangle in Figure 2.3 is deformed by an amount δt in time and δx in space

$$(c\Delta\tau)^2 = (c(\Delta t - \delta t))^2 - (\Delta x - \delta x)^2.$$

(2.26)

If we write $eA_0 = \gamma mc\delta t/\Delta t$ and $eA_1 = \gamma mc\delta x/\Delta t$ then we have the dispersion relation for an electron of charge e in an electromagnetic field (A_0, A_1)

$$(E - eA_0)^2 = (pc - eA_1)^2 + m^2c^4,$$

(2.27)

and the corresponding transfer matrix M is given by

$$M = \begin{pmatrix} \cos\theta_2 & \sin\theta_2 \\ \sin\theta_2 & -\cos\theta_2 \end{pmatrix} = \frac{1}{E - eA_0} \begin{pmatrix} mc^2 & pc - eA_1 \\ pc - eA_1 & -mc^2 \end{pmatrix}.$$

(2.28)

If as Eq. (2.22) we embed the causal net in a continuous 3-dimensional space we can replace p with the 3-dimensional momentum operator $\hat{\sigma}.\hat{p}$ and use the full vector form for the field (A, A_0) and write this as

$$\begin{pmatrix} mc^2 & c\hat{\sigma}.\hat{p} - eA \\ c\hat{\sigma}.\hat{p} - eA & -mc^2 \end{pmatrix} \Psi = (E - eA_0)\Psi.$$

(2.29)

Consider the case of non-relativistic motion in a weak field so that $E' = (E - eA_0)$ and $|E' - eA_0| \ll mc^2$. We can neglect the smaller component of the spinor and have, following [5], the Pauli equation for a non-relativistic spin-1/2 particle with $H = curl\ A$.

$$\left[\frac{(\hat{p} - eA)^2}{2m} + eA_0 - \frac{e\hbar}{2mc}(\hat{\sigma}.H) \right] \psi_1 = E'\psi_1$$

(2.30)

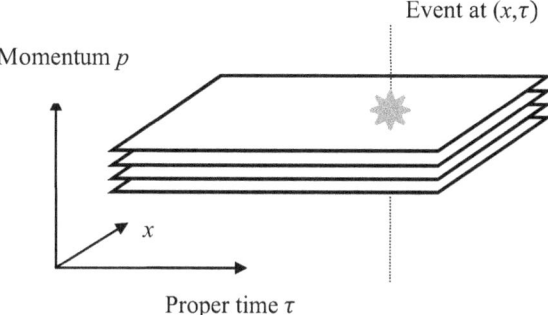

Figure 2.6: A stacked "deck" of causal nets for different momentum states.

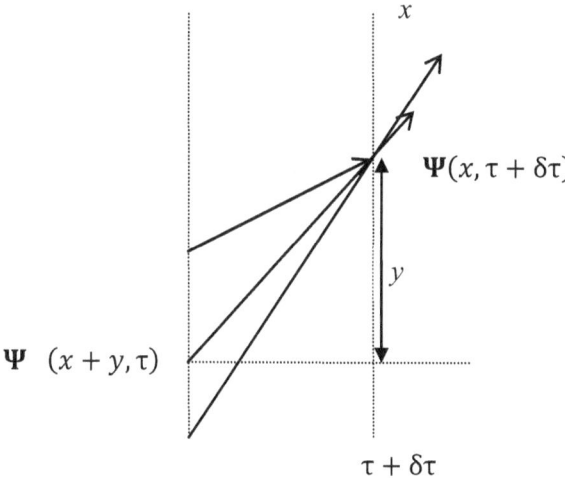

Figure 2.7: Causally connected paths traversing a single space-time point.

2.12 Momentum States and the Feynman Path Integral.

The causal net, denoted $\aleph$, we have considered is for a free particle with one specific momentum state. It is interesting to consider the more general case of a range of momentum states with each momentum state occupying an individual net $\aleph(p)$. This can be visualised in Figure 2.6 as a stacked "deck" of infinite causal nets. Importantly, due to the different net sizes (Eq. 2.1) the vertices of each net do not overlap. For a point in space-time (in practice this could be defined as an infinitesimal region of measurement) we can consider which events can causally act on a point at (x,t) from the past. To retain our definition of simultaneity as being defined by events that occur at the same proper time then it is a more consistent description to consider a point in proper time at $(x, \tau + \delta\tau)$ and the prior events that are causally connected from a earlier slice of proper time at τ which are given by *different* space points x from each momentum net $\aleph(p)$ (Fig. 2.7). This is equivalent to considering which possible particle paths pass through the point. To illustrate we shall consider only positive energy states but we can easily extend to include negative energy states. If we sum the different spinor components contributing to the overall probability amplitude at (x, τ) and include the change in phase over interval $\delta\tau$ from Eq. (2.6) we have

$$\Psi(x, \tau + \delta\tau) = \sum_{\substack{\text{causally}\\\text{connected}\\\text{events}}} \Psi(x + y, \tau)e^{-\frac{imc^2\delta\tau}{\hbar}}$$

$$(2.31)$$

here y is the relative space coordinate (Fig. 2.7). For one casual net with a free particle with a single momentum state Eq. (2.31) is trivial since velocity and probability are uniform across the net with only the phase varying with τ.

$$\Psi(x, \tau + \delta\tau) = M\, \Psi(x, \tau + \delta\tau) = \Psi(x + y, \tau)e^{-\frac{imc^2\delta\tau}{\hbar}}$$

$$(2.32)$$

providing a simple delta function propagator for proper time interval $\delta\tau$.

$$K(x, x + y; \tau + \delta\tau, \tau) = \delta(y)e^{-\frac{imc^2\delta\tau}{\hbar}}$$

(2.33)

However if there is a continuum of momentum nets by geometry the sum in Eq. (2.31) selects a single probability amplitude contribution from each net with momentum $p = my/\delta\tau$ for a given relative position y. Writing the relativistic infinitesimal action $S(\delta\tau) = -mc^2\delta\tau$ we can write Eq. (2.31) as an integration

$$\Psi\ (x, \tau + \delta\tau) = \int_{-\infty}^{\infty} \Psi_p(x + y, \tau)e^{\frac{iS(\delta\tau)}{\hbar}}\,dy$$

(2.34)

where Ψ_p denotes the spinor with momentum $p = my/\delta\tau$. If we consider the non-relativistic limit where one spinor component ψ_1 dominates and $\tau \to t$ we can use the semi-classical action

$$S_{cl} = \int_t^T \frac{m}{2}\left(\frac{dx}{dt}\right)^2 dt$$

(2.35)

and write this infinitesimal path integral in the limit of large time interval T to give the conventional Feynman path integral [33]

$$\psi_1(x, t + T) = \int_{-\infty}^{\infty} K_0(x, y; T)\,\psi_1(y, t)\,dy$$

(2.36)

where K_0 is the free particle propagator for the Schrödinger equation

$$K_0(x, y; T) = \left(\frac{m}{2\pi i\hbar T}\right)^{1/2} e^{imy^2/2\hbar T}$$

(2.37)

The causal net model is thus consistent with the quantum mechanical summation of paths and solutions to various problems such as slit diffraction using Feynman integrals as detailed in the classic text *Quantum Mechanics and Path Integrals* [8].

2.13 Non-Euclidean Space-Time and General Relativity.

Lastly, it is worth considering the case of non-Euclidean or curved space-time. Is our causal net model consistent with Einstein's general relativity? In curved space-time our elementary triangles (Fig. 2.3) comprising our causal net will become distorted and we can no longer apply Pythagoras' theorem to evaluate the space-time interval. If we embed our causal net in 4-dimensional space-time, then in the language of general relativity the space-time interval is given by the metric $g_{\mu\nu}$ so $ds^2 = g_{\mu\nu}dx^\mu dx^\nu$. Previously, we have considered the special case of the Minkowski metric $\eta_{\mu\nu}$ for flat space-time. General relativity considers Riemann spaces that have quadratic metric equations and are characterised as locally flat. This means that the first derivatives of the metric tensor are zero so for a small displacement in space Δx from a point x using Taylor expansion we have the metric

$$g_{\mu\nu}(x + \Delta x) = \eta_{\mu\nu} + \frac{1}{2}g_{\mu\nu,\rho\sigma}\Delta x^\rho \Delta x^\sigma.$$

(2.38)

So the change in the metric depends only on the second derivatives of the metric and is related to the *curvature* of the space-time. On the basis of Eq. (2.38) we might assume that variation to the scalar Minkowski action S (Eq. 2.34) would produce a correction S_g which, to be a scalar under general coordinate transformations, can only include second order derivatives of the metric. Mathematically, the simplest curvature scalar is the well known Ricci scalar $R = g^{\mu\nu}R_{\mu\nu}$ formed from the Ricci curvature tensor $R_{\mu\nu}$. We can postulate that the *simplest* additional action might be of the form

$$S_g = B \int R \, d^4x.$$

(2.39)

If we set the constant $B = -\varepsilon_0/16\pi Gc^4$, where ε_0 is a "density" and G Newton's gravitational constant then we have Einstein's *unimodular* gravity. If we further impose the density as $\varepsilon_0 = -\sqrt{-g}$ where $g = \det(g_{\mu\nu})$ then we recover the Einstein–Hilbert action

$$S_g = -\frac{1}{16\pi Gc^4}\int R \sqrt{-g} \, d^4x.$$

Now to balance the Einstein–Hilbert action, so that the overall action is stationary, we must add an additional term or Lagrangian. This, in general relativity, describes a matter field and is *defined* by setting $\delta S = 0$ to provide the characteristic stress-energy tensor $T_{\mu\nu}$

$$T_{\mu\nu} = -\frac{2\delta L_M}{\delta g^{\mu\nu}} + g_{\mu\nu} L_M,$$

(2.40)

and the famous Einstein field equations

$$R_{\mu\nu} - \frac{1}{2} g_{\mu\nu} R = \frac{8\pi G}{c^4} T_{\mu\nu}.$$

(2.41)

Thus the causal net model, although a microscopic theory, would appear to be consistent with the macroscopic theory of general relativity. The elementary triangles of our causal net model must be distorted to "tile" curved space-time between causally connected possible events.

Chapter 3

Discussion

3.1 Introduction.

In this Chapter we shall summarise and discuss the main theoretical results we have derived. It is informative to see how the causal net model provides an account of the major quantum phenomena including quantisation, the uncertainty principle and the common problems of diffraction and particle in a box. The causal net also provides an explanation for the transition from quantum to classical behaviour and the quantum measurement problem often described by the "collapse of the wavefunction". Non-local effects and the EPR paradox are also discussed. The causal net model is found to be consistent with the relative space-time of Leibniz and the flow of time determined by statistical measurement of events on the causal net. The emergence of fermions or Dirac particles from simple causal connections leads to the consideration that the laws of physics are perhaps emergent from the causal connectivity of the net. Lastly, no book on physics is complete without some wild speculation and we shall consider the possible role of a causal theory in particle physics, origin of fermion mass and gravitation and finally, routes to experimental verification of the causal net model.

3.2 A Brief Summary of the Results.

In Chapter 2 we showed that the Dirac equation could be derived from the fundamental concepts of causality and probability. By constructing the simplest possible causal net based upon the elementary Bravais lattice of possible events in space-time we have produced a framework that supports quantum mechanics and quantum mechanical statistics for a free particle.

Each possible event on the causal net represents a possible particle observation.

The causal net is consistent with the Reichenbach definition of simultaneity [18] and a relativistic form of the principle of common cause. Each possible event on the net has two possible common causes. Simultaneous possible events occur for equivalent proper time for the particle – that is the time that passes in the rest frame of the particle. The 1+1 dimensional net is aligned such that one dimension is the proper time τ and the second dimension the space dimension x aligned along the direction of velocity of the particle. The net is thus constructed from elementary triangles $(\Delta x, c\Delta t, c\Delta \tau)$ which are consistent, using Pythagoras' theorem, with the invariant space-time interval of special relativity. Actually, to retain causal invariance and not produce causal anomalies, we could argue that the simplest causal net *automatically* provides a finite maximal velocity (taken to be the speed of light c) and preserves the space-time interval. Thus perhaps, the axioms of special relativity are a product of the simplest equilibrium distribution of causally connected events.

Our treatment considers flat, Euclidean space-time but to maintain causality in bent or curved space we must distort the elementary space-time triangles and hence the space-time interval. This corresponds to a force or changing potential and has analogies to gravitation and general relativity which we discussed briefly in Section 2.13.

From the basic elementary triangles forming our net, once we have identified the space-time interval, we can derive the well known Lorentz transformation formulas. If we choose to scale the triangle by a factor m, the particle rest mass, we furthermore obtain Einstein's relativistic energy dispersion relation. Hence, mass m is an arbitrary, *ad-hoc* scaling factor. Thus possibly, the properties of velocity, mass, momentum and energy are observed, measurable quantities imposed by the observer on a simple causally connected series of events in space-time. The geometry of the net thus seems very important since it embodies all the physics of the problem and as we shall see the net angle is particularly important.

If we apply probability theory to our discrete lattice or net we can introduce connecting probabilities, as Reichenbach, between our net vertices. In fact by doing this our basic lattice of unconnected events becomes a true "net". If we apply probability conservation at each vertex or node we have what is called a Bayesian network. Probability conservation just means that if an event has occurred and a subsequent series of possible events can happen then their probabilities must sum to unity. No probability is lost or "radiated" away!

Now if we introduce a randomness or indeterminacy to events then an event on our net is linked to two possible future events. As noted this produces a correlation between events or an indeterminacy to particle motion (see Fig. 2.1). For the case of a free particle the particle can branch left or right at each vertex or discrete space-time point. This implies that, although on average the particle might move in a given direction, sometimes it moves in the opposite direction and a "zig-zagging" behaviour is observed on a microscopic scale (Fig. 2.2). This random movement is somewhat similar to the *Zitterbewegung* of a Dirac particle proposed by Schrödinger in 1930 [34].

At this point we introduced an assumption to describe the particle statistics. It is a simple, plausible assumption that the observed expected velocity of the particle is equal to its actual velocity. Consider how as an observer (in our rest frame) we would actually measure the expected velocity. Importantly, in our low dimensional example, we would assume that space and time are orthogonal (perpendicular) and measure a distance travelled in a certain time interval. An average of observations must converge towards the true velocity. Indeed, the true velocity must represent the drift in the diffusive behaviour of the particle. The particle can be easily shown to follow a diffusion process in proper time.

Introducing this assumption of average measured velocity in the observer frame and applying conservation of probability provides the branching probabilities at each node. For a free particle of one unique momentum value or state we can assume the velocity is the same across the entire net and the net extends in both time and space to infinity and has no boundary conditions. There is nothing wrong in this assumption since, as mentioned in Section 1.6.1, this is what quantum mechanics assumes for a free particle. This implies that the branching probabilities are the same at each vertex across the entire net and is the equilibrium condition. Interestingly, as we shall see, this suggests that quantum mechanics itself somehow represents an *equilibrium solution* in nature.

The probabilities at the net vertices can be represented by a matrix equation – in the jargon called a transfer matrix. The probabilities cross at each vertex and are coupled but do not interfere. They are as we shall see like wires carrying probability crossing at each vertex (see Section 3.8). The probabilities on the net are all positive real numbers. Does this allow any hidden symmetries to exist in our free particle scenario? The answer is yes. Any real number can be composed of other numbers and perhaps the simplest case is a dot-product of vectors or in a more conventional notation the modulus squared of a complex number. This allows the introduction of an underlying phase term which disappears when we compute the plane wave probabilities. If the phase is dependent on proper time then this is equivalent to what is known as global gauge invariance in physics or a so-called U(1) symmetry invariance. The phase variation in proper time can be mapped into the observer space-time using the relativistic invariance relations. It turns out that the concept of phase in proper time is the same as the first assumption used by de Broglie in his Nobel lecture on wave particle duality [4] and also is a base assumption of string theory.

Now the introduction of the phase in proper time is the simplest underlying symmetry and we can use this to form the probability amplitudes from the square roots of the probabilities. In this way taking the modulus-squared of the probability amplitudes recovers the probabilities on the net. For positive square roots of the probabilities the matrix equation that couples the probabilities has only one corresponding underlying transfer matrix *M*. This matrix *M* is unique, since it is the only matrix that provides conservation of both probability and probability amplitudes at the vertices. Amazingly, with this unique matrix if the probabilities at the vertices exist then the probability amplitudes *must* co-exist if there is global gauge invariance. The matrix *M* embodies strange symmetries. The matrix represents an improper rotation – a rotation through the net angle followed by an inversion – and the product of the matrix by itself provides the identity matrix.

The transfer matrix *M* for the probability amplitudes is the *exact* 1+1 dimension Dirac equation for the free particle. We have derived this with absolutely no approximations from first principles. The incoming (or outgoing) probability amplitudes at each vertex correspond to the exact two component spinor wavefunction for the 1+1 dimension Dirac equation.

Now are there any other possible solutions to our causal net? Previously we took the positive square root of both incoming (or outgoing) probability amplitudes at the vertices. These correspond to even solutions but an odd solution is available if one root is taken as negative. It turns out this provides a different fermion spin state. Thus the only two solutions to the causal net provide the two spin states of a fermion. A general state will be a superposition of these since they are energetically equivalent.

Additionally we have, as Dirac did, the possibility to consider a negative energy root solution to the relativistic dispersion relation. We don't really have any reason to do this except that it is a mathematical possibility. Dirac was forced to include them as solutions but we don't have to necessarily include them as valid. The two negative energy solutions correspond in fact to two negative energy spin states but are actually on our net negative velocity states. They have the same amplitudes as a net constructed with a negative momentum state but have a phase that evolves with time in the opposite direction. If we include these two negative solutions with their respective transfer matrices and combine with the two positive energy solutions we have the exact 4-component Dirac spinor. The 4 transfer matrices combined provide the full, well known Dirac equation.

3.3 The Transition to Classical Behaviour.

A major issue in quantum mechanics is the conceptual bridge between the classical and the quantum world as emphasized by the notorious "Schrödinger cat" problem [35]. In the causal net model this has a ready explanation since a cat is a massive object and not subject to any quantum mechanical effects – even though before we open the box we do not know whether the cat is alive or dead the cat certainly knows that he is alive! We find that on the causal net the transition from quantum to classical behaviour occurs for massive objects

For massive objects since $\Delta x = h/2p$ from the uncertainty relations when p is large – true for high mass or virtually any velocity – then the net size Δx is small and uncertainty in x is small relative to the size of the object. The object can be well localized or resolved on the net and is effectively non quantum although of course it can be relativistic.

In recent years a theory of decoherence has been developed whereby phase information of the quantum wavefunction is destroyed through various proposed processes leading to classical behaviour [36]. This process is usually described by introducing a spatial or temporal coherence length for different environments and damping the cross terms in the density matrix. In our model a disruption of causal links by some form of environmental "noise" or interaction with a classical object would "break" the probability distribution or "collapse" the net. This effect would be similar to an actual event or act of measurement. Quantum behaviour would then only be readily observable on coherent "islands" within the overall classical world characterized by coherence lengths. As we mentioned above, quantum objects in the causal net model are low mass with finite velocity and massive objects behave classically. The mechanism for decoherence may be therefore an interaction which constitutes a brief coupling to a massive classical object.

3.4 Quantisation, Particle in a Box and Diffraction.

It is worth emphasising that the full conventional Dirac equation arises only when we impose probability amplitudes on the causal net. In contrast the quantisation is automatically provided by the net and does not rely on the full formalism of quantum mechanics in terms of the wavefunction. To illustrate this consider the relativistic particle in the (1-dimensional) box problem shown in Figure 3.1. For a potential well of depth V and width L bound states exist for a "forbidden zone" given by imaginary momentum states in the potential region. Now from the causal net model we can consider there to be an integer number n of net vertices to be contained in the well. This provides the quantization condition $n = L\Delta x$ giving quantised momentum states $p = 2nL/h$. This corresponds to the solution of the 1+1 dimensional Dirac equation [31] and reduces to the Schrödinger particle in the box problem in the non relativistic limit. Thus for the particle in say its $n = 2$ state then there are 2 space events on the net that can occur at the same time with equal probability.

In conventional quantum mechanics the bound state problem is treated as a superposition of two plane waves with opposite momentum states and a reflective boundary condition applied at the edge of the potential well. In our model we could treat the problem as a superposition of two causal nets as Section 2.12 although as we have seen this is not necessary to realise that any bound state is quantised into maxima and minima since this is given by the discrete nature of the net, as described above.

Since we have demonstrated that a "stack" of causal nets provides the Feynman path integral (Section 2.11) we can use this well known technique to solve other quantum problems such as diffraction from slits. If a single net is denoted by $\aleph$ then we must consider the set of $\aleph(p)$ nets of all possible p momentum states applicable to the problem. Following Feynman, as detailed in his famous book *Quantum Mechanics and Path Integrals* [8], for a single slit all possible trajectories of the particle which can travel from the source and through the slit(s) must be considered. The sum over paths provides well known Fresnel integrals which provide the observed diffraction phenomena at the detector. To facilitate the computation the notion of particle kernels or propagators are used (Section 2.12). In the language of our causal nets we must consider all causal chains that causally link the source and the detector that are not limited by the slit apparatus. We do not know which causal path was followed to reach the detector but upon final measurement we could, if necessary, compute using Bayesian analysis (see Section 3.6) the range of possible paths the particle could have followed.

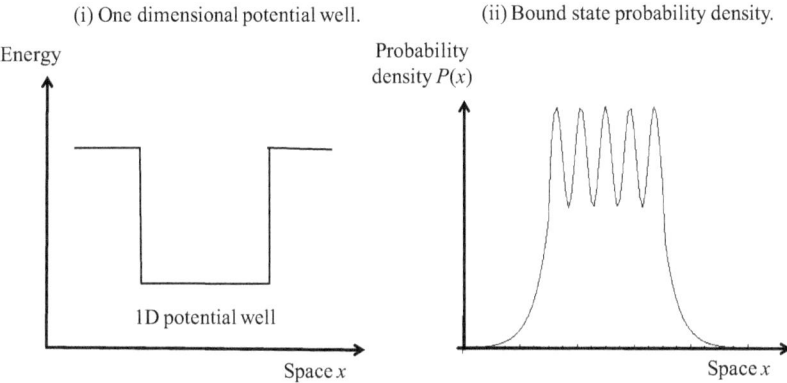

Figure 3.1: The quantum particle in the box.

3.5 The Uncertainty Principle.

In Section 2.5 we saw that $p\Delta x$ is a constant η for a valid lattice and, for the particle to obey the simplest form of the principle of least action, that the sum $\sum p\Delta x$ is the same for all paths on the causal net. Setting this constant η to be Planck's constant h is an arbitrary decision but provides the de Broglie relation $\lambda p = h$ with lattice constant $\Delta x = \lambda/2$. The discretisation of the net then provides a Heisenberg like relation $\Delta p \Delta x \sim h/2$. The exact published definition of the Heisenberg relation can differ and the numerical variable can sometimes include a factor of $1/\pi$ depending on the experimental situation considered.

Now a minimum discretisation of space-time is perhaps not so far removed from the early ideas of Heisenberg himself. The *uncertainty principle* [29] was recognized by Heisenberg himself in 1927 as being possibly caused by a discontinuous space-time system: "When one admits that discontinuities are somehow typical of processes that take place in small regions and in short times, then a contradiction between concepts of "position" and "velocity" is quite plausible". By considering motion to be a series of points of finite separation he continued: "In this case it is clearly meaningless to speak about one velocity at one position (1) because one velocity can only be defined by two positions and (2), conversely, because any one point is associated with two velocities." Although Heisenberg did not go as far as to propose a discretised space-time we have shown that the causal net model can provide the uncertainty relations and the de Broglie relation. These are only seen to be relevant for quantum mechanical objects though since for massive objects matter behaves classically (Section 3.3).

3.6 The Quantum Measurement Problem.

The causal net is a net of probabilities and a measurement provides through Bayesian statistics a re-evaluation of these probabilities after a measurement – equivalent to "collapse of the wavefunction". The causal net provides a net of possibilities in space-time describing the position and momentum of the particle.

To illustrate, assume as Chapter 2 the probability of finding a particle at vertex labelled (i,j) on a measurement is given by $\Omega_{i,j}$. Thus a statistical measurement will provide through Bayesian analysis a change in all the probabilities on the net for the current time and all the previous times. The mysterious "collapse of the wavefunction" just corresponds to a statistical measurement and the change or refinement in all the retrospective probabilities. This is easily seen for the case of a single particle. Consider two different space locations at the same proper time. If for example the probability of finding the particle at $\Omega_{1,2}$ is the same as $\Omega_{2,2}$ before the measurement and we perform a measurement at vertex (1,2) and the answer is "yes we have found a particle" then $\Omega_{1,2} = 1$ and $\Omega_{2,2} = 0$. Instantaneously, from Bayes theorem and basic probability theory, the probabilities earlier in time will change: $\Omega_{1,1}$ will increase and the probability $\Omega_{2,1}$ will decrease.

The above approach has reduced the general quantum mechanical measurement process to one of causal statistics of an ensemble of events. Determination or measurement of the actual state of a vertex comprises a statistical measurement which is fully consistent with all the probabilities on the net. If we consider a single particle similar to the Aspect experiments then when the particle is measured in a particular location there is no true "collapse of the wavefunction" but merely a statistical observation yes or no. From this observation Bayesian analysis can be applied to compute the retrospective or historic probabilities and statistics of earlier vertices but these are all perfectly consistent with probability theory.

So-called "action at a distance" is just a natural consequence of measuring and statistically determining the state of the particle of the system. There is no faster than light transmission of information as many people claim. If my colleagues think I might have gone on holiday to France or Mauritius and I send them a postcard from Mauritius then when they receive it they can instantaneously re-evaluate the probability of holidaying in Mauritius is one and the probability of being on holiday in France is zero – there is no faster than light communication between France and Mauritius!

3.7 Interpretation of Relative Space-Time.

The casual net model provides a relational space time similar to that envisaged by Leibniz rather than the absolute space time of Newton. Space and time are constructed from the causal relations among events and the measured distance between them. For us the causal net is comprised of elementary *possible* events and a measured event is an observed particle.

This nature of discrete, indivisible events is also similar to the concept of Leibniz's monads described in Section 1.3.3. Interpreting in an extreme sense Leibniz, his view was that there is no unoccupied space and time between real events: "I do not say that matter and space are the same thing; I only say that there is no space where there is no matter, and that space itself is not an absolute reality. Space and matter differ as time and change [movement]. However, these things, though different are inseparable. But it does not at all follow that matter is eternal and necessary, unless we suppose space is eternal and necessary – an altogether unfounded supposition."

Attempts have been made to fuse relativity and indeterminism by philosophers such as Belnap and others [37]. Like us they assume that a theory of branching space-time can be built on the primitives of a set of "possible point events" and the causal relations between them. They have not however provided a link between their theories and "real physics" and hopefully the causal net model we have developed helps fill this gap.

We could consider the net and the set of all nets exist only as a mathematical description (as say Pythagoras' theorem) or a computing device for possibilities in the same way that a particle on the net is probabilistic. Only when an event or measurement occurs then the region of space-time has reality and an existence and until then it remains a possibility represented by probability. As with the vague Copenhagen interpretation we cannot say anything about the particles trajectory when it is not observed. As one friend of mine (a renowned theoretical physicist) once said to me on a chairlift in Jackson Hole "anybody who asks if a quantum particle has local reality between observations is asking for a punch in the face!"

In our interpretation, between observations particles have their own local reality and "exist" in their own stationary reference frame since probability is conserved and the particle is not moving anywhere. The precise position the particle occupies in space-time in other inertial frames is unknown. Determining the location of the particle in the observer frame temporarily collapses the causal net for that particular reference frame. This agrees with the wave-particle duality of nature – that particles move on the probabilistic causal net and appear as waves when they are not observed. However, when we measure them we have and actual event and localise and find a particle. Special relativity showed that there is no universal time but we have replaced it with each particle having its own invariant proper time or "personal" time – a kind of "universal personal time". Obviously this is an essentially realist interpretation of space-time – that both space and time exist outside the human mind.

3.8 A Physical Analogy – Crossing Wires.

We can perhaps consider a common physical analogy for the causal net. Combining the probability amplitudes (Eq. 2.12) in a linear manner is analogous to combining alternating or AC currents in an electrical circuit linearly to preserve phase before calculating power emitted. For the case of a free particle we previously noted that $\mathcal{P}_{1,1} = \mathcal{P}_{2,2}$ and $\mathcal{P}_{1,2} = \mathcal{P}_{2,1}$ so the causal net is analogous to electrical currents "crossing" in wires at each vertex and the probability corresponding to the total instantaneous power (Fig. 3.2).

If we write each individual spinor component as a current with complex phase, varying in proper time given by Eq. (2.6), then the sum of the real measurable *apparent power* from each crossing current is equivalent to the probability at the vertex. The apparent power is the modulus squared of the current multiplied by a wire resistance that is a scaling constant. This analogy is quite remarkable since we are usually taught to think in terms of probability currents but in fact if we think instead of the probability amplitudes being complex electrical currents and the probability being the radiated power the physical analogy is much more precise. Also interestingly the power in an electrical circuit is a conserved quantity (power in = power out) and this is also true of probability. To think of the Dirac equation being represented as a bunch of crossing electrical wires is very bizarre!

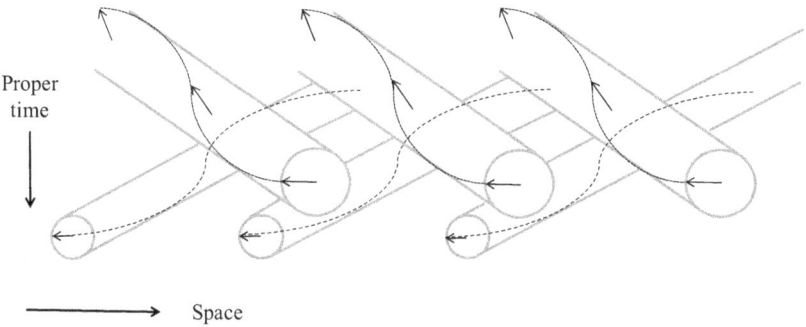

Figure 3.2: The causal net is analogous to crossing wires with AC current.

3.9 The Flow of Time.

Many different views on the nature and flow of time have been presented over the years. Newton stated in his *Principia* that "Absolute, true and mathematical time, of itself, and from its own nature, flows equably without relation to anything external, and by another name is called duration." On the basis of relativity Einstein stated that "the distinction between past, present and future is only an illusion, albeit a stubborn one." Others have taken more radical views such as Hoyle who went further and claimed that "all moments of time exist together". Popular these days (amongst trendy cosmologists) is that the macroscopic arrow of time is set by the Big Bang and the evolution of the universe.

In our model of relative time the fundamental concept is that the causality of events must be preserved. The only "absolute" time is the proper time for each particle. This is consistent with special relativity where causality is retained between events but time intervals between events for observers in different inertial frames will vary. In relativity the fact that time can be treated almost as another dimension has lead to the predominant view that time (multiplied by a velocity c) is somehow equivalent to space. However, the geometrical construction of the causal net demonstrates that this is misleading since although time and space intervals can be interchanged, depending on the inertial frame, proper time and space have markedly different characteristics.

Interesting the causal net with the principle of the common cause has other possible implications relating to the flow of time and irreversibility. Reichenbach [18] argued that second law of thermodynamics could be derived from application of the principle to statistical thermodynamics. On the flow of time he writes: "Neither the laws of [classical] mechanics nor mechanical observables give us the direction of time, unless such a direction has been defined previously by reference to some irreversible process. For instance, if the velocity of a body is regarded as observable, its direction must be ascertained by comparison with some temporally directed process such as the time of psychological experience, which is derived from the irreversible processes of the human organism." In our causal net theory of quantum mechanics actual statistical measurement of vertex states define such irreversible processes which can provide a direction to time.

3.10   Emergent Particles and Laws.

From the basis of simple casual connections between elementary events we have constructed a model where the Dirac equation and the fermion particles it describes are seemingly emergent properties. Does the causal net automatically imply a quasiparticle such as a fermion? In an inertial frame an observer will view an ordered series of events in space-time as an entity behaving as either a wave or a particle, depending on how the experimental measuring setup is conceived, and thus exhibits wave-particle duality exactly as proposed by de Broglie. Geometrical quantities of the causal net correspond to measurable physical qualities: mass (scaling factor), momentum and energy (net angle and geometry) and acceleration and forces (change of net angle). This is illustrated in Table 3.1. The global gauge symmetry of the net provides quantum phase and the other degenerate solutions arising from the symmetries of the net provide the Dirac spin and negative energy states. In addition the discretisation of the net provides an analogy with the Heisenberg uncertainty principle and a "stack" of nets provides, consistent with the Feynman path integral approach, quantum phenomenon such as diffraction. None of these emergent aspects of our net would have been apparent from our simple starting point of an equilibrium distribution of ordered events.

Table 3.1: Physical analogies of the causal net.

Causal net parameters	Physical parameter and concepts
scaling factor	mass
net angle	velocity
net triangle geometry (increased hypotenuse)	energy
net triangle geometry (increased opposite side length)	momentum
change in net geometry	force or changing potential
gauge invariance (global)	phase
square root of probabilities – odd and even solutions	spin
square root of dispersion relation – odd and even solutions	positive and negative energy states
net discretisation	uncertainty principle
"stack" of nets	different momentum states and Feynman path integral
non-Euclidean geometry	Hilbert action, general relativity

We can see that in nature virtually every symmetry of the causal net is exploited. By imposing extra symmetries different physical phenomena can be accommodated. For example, introducing a local U(2) gauge invariance, say a rotation, to the wavefunction provides an extra term in the net Hamiltonian that corresponds to an electromagnetic potential. Just by measuring events with increasing net angle in space and time we would, as an observer, infer (in correspondence with quantum theory) that an electromagnetic potential exists acting on a Dirac particle such as an electron.

In addition we have demonstrated the possibility of computing quantum mechanics using only real, positive probabilities using the causal net. The use of complex numbers in quantum mechanics is usually taken as mysterious but in fact everything measurable in quantum mechanics is a real number and can be computed from a vector (as Section 2.5) and matrix representation where only real elements are required.

The Dirac equation on the discretised causal net does not require spatial or time derivatives and we only introduce these to correspond to the more familiar continuous space-time case of quantum mechanics. Thus the net Dirac equations are much simpler and possibly more fundamental than the actual differential Dirac and Schrödinger equations. Quantum mechanics is based on the concept of differential operators and indeed Dirac demonstrated that quantum mechanics could be built from the ground up from these operators. These operators are differential equations that act on the wavefunction to provide a particular measurable quantity such as position, momentum, energy, etc. All the usual quantum operators are compatible with our causal net theory and also their Foldy–Wouthuysen analogues (Section 2.8). Is it possible that these operators are just a human concept allowing us to more easily perform quantum calculations, using the continuous methods of calculus, and have no true fundamental basis? In our model the causal net is similar to a Dirac state vector and so are the Dirac operators just mathematical tricks to extract information from the causal net?

If we take the above discussion a bit further we might speculate that most of the equations and laws of physics that we know and love are "invented", based on differential operators and the methods of calculus applied to a fundamentally discrete space-time system comprised of causally connected events. This presents rather a depressing view of physics as a "hollow" theory where all we are doing as scientists is "fitting" or "constructing" a theory of physics, with the minimum number of parameters, to a series of events that we observe in space-time. For the most part our theories will then represent equilibrium conditions or distributions of events that can be repeated and measured under laboratory conditions. This is not perhaps a totally new or radical notion since Eddington in his *Philosophy of Physical Science* (1939) [38] proposed that physics might be subjective and our perceptions are influenced by our human characteristics. He viewed much of physics as invented by humans rather than truly discovered.

It is interesting that we have considered only the simplest configuration of causal net – a Bravais lattice – and this reproduces the Dirac equation. Different topologies of causal net might give different results and it would be worthwhile exploring these. We also note that the topology of the causal net can be mapped directly onto an ordered spin-ice lattice in solid state physics. Recently magnetic monopoles have been observed as breaking of the "ice rules" in frustrated magnets such as spin-ice ($Di_2Ti_2O_7$) and magnetic charge and current observed [24].

3.11 Non-Locality and the EPR Paradox.

It is widely accepted that quantum mechanics cannot be developed from a basic application of Reichenbach's principle of common cause with a *single* conjunctive fork since general quantum mechanical statistics violate the principle. The principle of common cause involving a single conjunctive fork can even actually be used to formulate the well known Bell's theorem [15], which motivated the famous experiments by Alain Aspect [14] to exclude the possibilities of certain types of hidden variables. Here we have considered a modified framework where a complete causal network of possible events is comprised of conjunctive forks such that each possible event has *two* effective local common causes or screening factors. Adjacent possible events on the net that are simultaneous are thus considered to share a common cause. It appears that, at least in the simple case we consider, this allows our common cause principle – based on the simultaneity of neighbouring *possible* events – to be applied consistently with quantum statistics.

In many ways this statistical causal net model is a non-local hidden variable theory – the hidden variable could be considered to be the actual net. Bell's theorem applies to local hidden variable theories and since our theory is a non-local hidden variable theory which exactly derives the Dirac equation it should always pass Bell's tests and give the exact quantum mechanical result. The causal net model is perhaps also consistent with the famous Einstein, Podolsky and Rosen (EPR) paper [39] that the "description of reality as given by a wave function is not complete" since the wavefunction is not complete without the connectivity of the causal net.

Our analysis of the superposition of nets in Section 2.12 could be possibly extended to an EPR type spin experiment with parallel settings. The experimental apparatus is shown schematically in Fig. 3.3 below. Two spin ½ particles are simultaneously emitted in opposite directions from a source in a singlet (or antiparallel) state. The particles are later detected along three different chosen spatial axes using magnetic fields in the left and right wings of the apparatus.

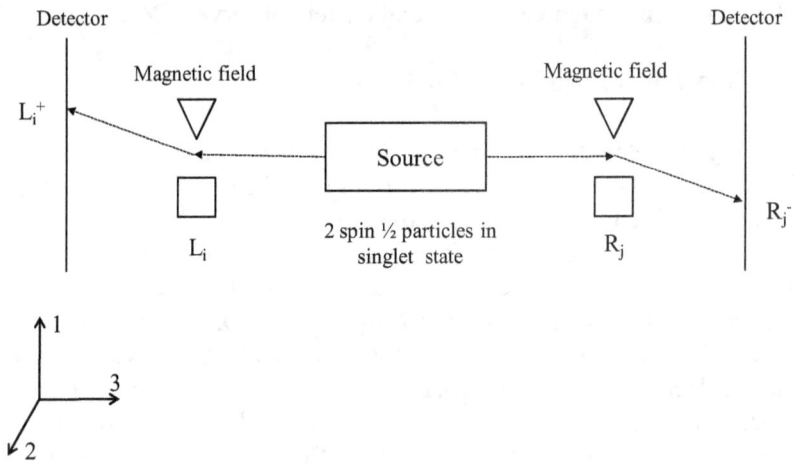

Figure 3.3: EPR experimental setup (from [40]).

Following the notation of Grasshof [40], if the measurement apparatus is set to measure the spin direction in the left and right wings $L_i(R_j)$ $i \in$ {1,2,3} $j \in$ {1,2,3} and $L_i^a(R_j^b)$ symbolizes the spin event type $a \in \{+, -\}$, $b \in \{+, -\}$ then the experiment measures $P(L_i^a. R_j^b | L_i. R_j)$. However for two particles there exists a set of possible pairs of causal nets $\aleph(L_l. R_m)$ comprising all possible experimental setups l and m for each particle. If we choose to measure the combination $\aleph(L_i. R_j)$ then we select the relevant pair of nets. This selection is mutually exclusive of all the other possible pairs of nets existing in the set. For perfect correlation $i = j$ and we can see that $P(L_i^+ | L_i. R_j) = 1/2$ and due to the initial antiparallel spin arrangement the other net must provide $P(R_i^- | L_i. R_j. L_i^+) = 1$.

For imperfect correlation we must consider probabilities such as

$$P(L_i^-.R_j^+|L_i.R_j) = P(R_j^+|L_i.R_j.L_i^-)P(L_i^-|L_i.R_j)$$
$$= f(\varphi_{ij})/2.$$

Due to mutual exclusivity it does not matter the order we measure L_i or R_j since determination of one state implies the other. The function $f(\varphi_{ij})$ where φ_{ij} is the angle between i and j can be considered to be a projection of a causal net corresponding to projecting R_i onto a different spatial axis j to give R_j and has a form $f(\varphi_{ij}) = cos^2\varphi_{ij}$. This result corresponds to quantum mechanical statistics.

Thus importantly, it would seem that there exists a full set of possible causal nets ℵ between physical observations. This set of causal nets is associated with the entire wavefunction of the system. Combining many simple causal nets together we can form a sort of "super" causal net that accommodates not only all possible position and momentum states but also various possible experimental arrangements and corresponds to the entire wavefunction. On a physical observation or measurement one member (or subset in the case of several particles) of the set of all possible nets ℵ is selected through mutual exclusivity based on the experimental arrangement and detection of the particle(s). This is somewhat similar to the theory of de Broglie [41] where a kind of "many worlds" quantum state exists in the microscopic world but in the macroscopic world only one member of the ensemble is statistically realised when a particular measurement is specified. In our model the mechanism for selection of one member of the set of causal nets and measurement of a particle on the net is through *mutual exclusivity* – that is the particle can exist at a particular location on the net and only one net fits the experimental conditions. The existence of one net per fermion state is consistent with the Pauli exclusion principle.

Lastly, we shall finish with a brief comment by John Bell himself in 1986: "because of the EPR experiments ... I want to say there is a real causal sequence which is defined in the aether." Bell did not imply that the original notion of the aether (Section 1.4.2) should be resurrected but "behind the apparent Lorentz invariance of the [quantum] phenomena, there is a deeper [causal] level which is not Lorentz invariant."

3.12 LHC, Mass, Kaluza and Black Holes.

Amidst much fanfare, hype and publicity the LHC or Large Hadron Collider at CERN is starting up this year. Will it find the Higgs particle or anything else interesting? Hopefully, the particle physicists will discover something new to pacify all us tax payers who contributed a fortune to purchasing all those super-conducting magnets. Years ago I worked briefly on the also super expensive LEP (Large Electron Positron) collider at CERN – the predecessor of LHC which was housed in the same 27km tunnel. It was a very interesting experience but with several hundred physicists working on each of the four detectors and everything planned out years in advance it was hardly a creative individual exercise. When I asked my boss what physics we were meant to be testing he replied half joking: "I can't really remember. It's in one of those old books there" and he pointed towards a dusty pile of design specs. At least I was there for the first few particle collisions, the opening party and some climbing at Chamonix. Eventually I fled to the Institut Laue Langevin (ILL) neutron facility in Grenoble, France where you could design, perform and analyse your own experiments which was intellectually much more fun. However, at CERN I pondered the highly successful standard model of particle physics. Although it must be considered an extraordinary achievement and although it possessed beautiful symmetry in places it seemed to me somewhat "cobbled" together. Its primary function seemed to be a means of categorising the "animals" in the particle zoo. Is it altogether surprising that smashing particles together at higher and higher energies produces certain symmetrical groups of particles? Under these slightly artificial fireball conditions maybe we are just witnessing different excited states of something more fundamental? Are these plethora of particles just emergent properties of our causal net framework?

It is currently indisputable, on the basis of experimental evidence, that there exists 12 fundamental, probably indivisible fermions. Is it a coincidence that the conventional Dirac equation is actually not unique in that in 3+1 dimensions the formulation of the matrices can provide 12 degenerate (and only 12!) different variations or 6 pairs of Dirac equation? Do these represent the 3 lepton generations and the 3 quark generations and represent the 12 different fermion fields?

Figure 3.4: The world's largest superconductor solenoid used on the LHC at CERN. ©
CORBIS

Currently our framework tells us nothing about mass and charge that
characterise the different fermions. Mass in our model is just an arbitrary
variable and it would need some sort of spectrum which coincided with
the fermion masses to extend these ideas. This is somewhat similar to
requiring a fifth dimension like the Kaluza–Klein tower [42] to provide a
spectrum of mass states. Kaluza came up with the idea of a fifth
compactified dimension that he sent to Einstein to help him get published.
Einstein thought it was a good idea so delayed Kaluza and worked on it
himself and only helped Kaluza when his own work on the subject could
be published at the same time! By this time the poor Kaluza had abandoned
physics altogether. A nice story of small-minded academic rivalry by great
minds... Anyhow the "miracle" of introducing Kaluza and later Klein's
fifth dimension was the apparent unification of relativity and
electromagnetism. Depending how we formulate the fifth dimension the
particle mass can be related to the extra coordinate introduced and electric
charge the fifth dimension momentum. Charge can be apparently

quantised if boundary conditions are placed on the extra dimension. Unfortunately, although offering tantalising possibilities and demonstrating that variables such as mass and charge could possibly arise out of the geometry of extra dimensions, the Kaluza-Klein model does not, on its simplistic assumptions, accurately model any extra actual observables we can measure (such as the fermion masses or electric charge) in the real world. Campbell's theorem demonstrated that our 4D world could be embedded in a universe containing any number of extra dimensions and a 5D world could contain an extra time or space dimension. Research on Space-Time-Matter (STM) theory [44] (my favourite approach) and increasing the number of dimensions (supergravity and super strings) has not, as we know, produced a suitable theory. It is not in my opinion that these are "dead ends", as some people would suggest, but really a reflection that with increasing dimensions there are a huge number of permutations of possible models to investigate – like "finding a needle in a haystack". The difficulty is compounded by the fact that from our 4D vantage point we can only experimentally observe the "tops of the icebergs".

In our causal net model it is questionable whether extra dimensions are really dimensions or merely extra variables or degrees of freedom. If an extra dimension is time-like, and the 5D action is taken as zero (a standard assumption) then this leads to oscillatory solutions in proper time. This is analogous to the phase in the wavefunction (in Eq. 2.10) so perhaps the fact that an event can contain phase information that is multi-valued at each space-time point is an obvious indication of a closed time-like dimension (or internal degree of freedom). Kaluza–Klein theory indicated that electric charge should be similarly quantised from a time-like dimension and should provide an effective mass in our observed 4D universe that depends on the electric charge. Allowing mass to be determined from a space-like dimension would possibly provide an exponential spectrum of fermion masses. Experimentally the fact that observed fermion masses follow an exponential hierarchy with variations (particularly for the quarks) depending on charge is I feel encouraging (Fig. 3.5). Perhaps these so-called extra dimensions or variables involve a combination or superposition of time-like and space-like degrees of freedom and it is possible to construct a causal net theory to provide this.

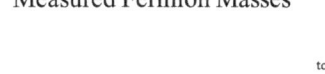

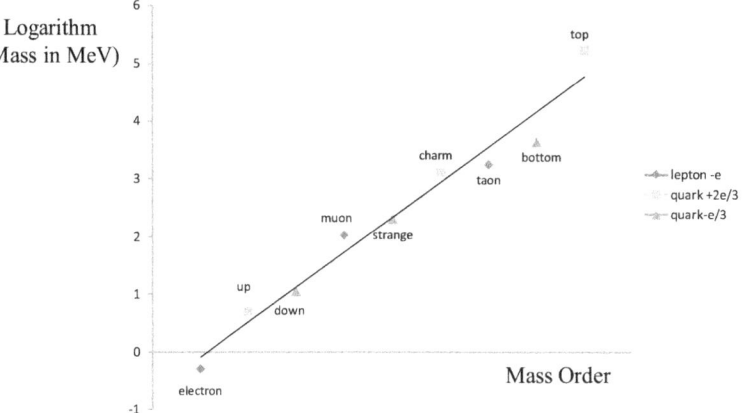

Figure 3.5: The measured fermion masses have an approximate "space-like" dimension exponential hierarchy varying also perhaps with a Kaluza "time-like" dimension electric charge.

Returning to general relativity, we remarked in Section 2.13 that the invariant interval in our causal net model provides a link between quantum mechanics and general relativity. Modifying the phase or action on the net by the simplest scalar curvature – the Ricci scalar – provides the Einstein–Hilbert action. Thus the causal net model, although a microscopic theory, would appear to be consistent with the macroscopic theory of general relativity. The elementary triangles of our causal net model must be distorted to "tile" curved space-time between causally connected possible events. So, is this a chicken before the egg problem – what comes first geometry or causality? However, recalling Reichenbach's argument that causality determines the geometry of space (in Section 1.7.2) it would seem that it might be possible to produce a theory of quantum space-time based on causality generating curvature that appears as an effective gravitational force or acceleration. However, it would seem that such an approach, based on discrete space-time, would possibly never achieve a singularity desired by black-hole fans.

Einstein speculated on the possibility of a causal theory underlying general relativity and in 1916 he wrote "Causal set theory arises by combining discreteness and causality to create a substance that can be the basis of a theory of quantum gravity. Spacetime is thereby replaced by a vast assembly of discrete "elements" organised by means of "relations" between them...None of the continuum attributes of spacetime, neither metric, topology nor differentiable structure are retained, but emerge it is hoped as approximate concepts at macroscopic scales." Interestingly, Einstein predicted the emergence of the known physical laws from a discrete causal structure. In recent years Rafael Sorkin and others have explored a causal set approach to quantum gravity [43] but our causal net is Lorentz invariant without assuming their "Poisson sprinkling".

3.13   Conclusion.

From our causal net model we have argued that what we call "physics", with all its well known laws and differential equations, is ultimately just a humanistic invention we have formed from empirical observation of the emergent properties of nature. Nature, from fermions to the termites and their mounds, is then emergent from a series of causally connected events that comprise relative space-time. Following P.W. Anderson (Section 1.8), emergent phenomena at different length and time-scales provides the everyday world we experience.

The causal net model suggests there is a new, important "layer" beneath our currently accepted fundamental equations of physics. Now, I strongly believe that all theories of physics should be experimentally testable. How is the causal net model possibly experimentally verifiable? One obvious route may be to recognise that quantum mechanics is an equilibrium condition and that the causal net approach allows us to consider what happens in non equilibrium conditions. For example, quantum mechanics describes clearly an equilibrium condition for a particle in a box (Section 3.4) but what happens immediately after a particle is placed in a box or after it is observed at a particular location within the box? How is a new equilibrium condition attained? The discrete causal net model makes strong predictions for such non equilibrium conditions, particularly for relativistic particles where the second component of the spinor cannot be neglected. Perhaps it is possible to construct clever non equilibrium experiments to verify the causal net model? Verification of the causal net model would allow us to view so-called "fundamental" equations, such as the Dirac equation, as merely emergent in a similar way to Newton's equations being emergent from quantum mechanics.

Bibliography

1. Kant I. (1787) *Critique of Pure Reason*.
2. Clarke S. (1717) *A Collection of Papers which passed between the late learned Mr Leibniz and Dr Clarke in the years 1715 and 1716. Relating to the Principle of Natural Philosophy and Religion*.
3. Einstein A. (1905) *Zur Electrodynamik bewegler korper*, Annalen der Physik 17, p. 891.
4. De Broglie L. Ann. Phys. (Paris) 3, p. 22 (1925) and Nobel lecture (1929) *The Wave Nature of the Electron*.
5. Davydov A.S. (1965) *Quantum Mechanics* (Pergamon Press).
6. Dirac P.A.M. (1928) *The Quantum Theory of the Electron*, Proc. Royal Soc. A117.
7. Dirac P.A.M. (1930) *The Principles of Quantum Mechanics* (Oxford).
8. Feynman R.P. and Hibbs A.R. (1965) *Quantum Mechanics and Path Integrals* (McGraw-Hill).
9. Kauffman L. H. and Noyes H. P. (1996) *Discrete Physics and the Dirac Equation*, SLAC–PUB–7115.
10. Gaveau B., Jacobson T., Kac M., Schulman L.S. (1984) *Relativistic Extension of the Analogy between Quantum Mechanics and Brownian Motion*, Phys. Rev. Lett. 53, p. 419.
11. Coddens G. (1995) *A Remark on the Dirac Equation*, Found. Physics Lett., 8, p. 3.
12. Bialynicki-Birula I. (1994) *Weyl, Dirac, and Maxwell Equations on a Lattice as Unitary Cellular Automata*, Phys. Rev. D, 49, p. 12.
13. Schotte K.D., Iwabuchi S. and Truong T.T. (1985) *Ice Models and a Lattice Version of the Dirac Equation*, Z. Phys. B 60, p. 255.
14. Aspect A., Dalibard J., Roger G. (1982) *Experimental Test of Bell's Inequalities using Time-Varying Analyzers*, Phys. Rev. Lett. 49, p. 1804-1807.
15. Bell J. (1966) *On the Problem of Hidden Variables in Quantum Mechanics*, Rev. Mod. Phys., 38, p. 447-52.
16. Rae A. (2004) *Quantum Physics* (Cambridge University Press).
17. Bohm. D. (1952) *A Suggested Interpretation of the Quantum Theory in Terms of "Hidden Variables"*, Phys. Rev. 85, p. 166.
18. Reichenbach H. (1956) *The Direction of Time* (University California Press).

19. Salmon M.H. *et al* (1999) *Introduction to the Philosophy of Science* (Hackett Publishing Company).
20. Penrose O. and Percival I. C. (1962) *The Direction of Time*, Proc. Phys. Soc., 79, p. 605.
21. Malament D. (1977) *Causal Theories of Time and the Conventionality of Simultaneity*, Nous, 11, p. 293.
22. Anderson P.W. (1972) *More is Different*, Science, 177, p. 393.
23. Laughlin R.B. (2005) *A Different Universe*, (Basic Books).
24. Bramwell S.T., Giblin S.R., Calder S., Aldus R., Prabhakaran D., Fennell T., (2009) *Measurement of the Charge and Current of Magnetic Monopoles in Spin Ice,* Nature 461 956.
25. Wolfram S. (2002) *A New Kind of Science* (Wolfram Media).
26. Bostrom N. (2003) *Are You Living in a Computer Simulation?* Phil. Quart. 53, p. 243.
27. Jaynes E.T. (2003) *Probability Theory: The Logic of Science*, 2nd Ed. (Cambridge University Press).
28. Rindler W. (1982) *Introduction to Special Relativity* (Oxford).
29. Heisenberg W. (1927) *Uber den Anschaulichen Inhalt der Quantentheoretischen Kinematik und Mechanik*, Zeitschrift fur Physik, 43, p. 172-198.
30. Schurmann T. and Hoffmann I. (2009) *A Closer Look at the Uncertainty relation of Position and Momentum*, Found. Phys. 39, p. 958.
31. Coulter B. and Adler C. (1971) *The Relativistic One Dimensional Square Potential*, AJP 39, p. 305.
32. Foldy L. and Wouthuysen S. (1950) *On the Dirac Theory of Spin ½ Particles and its Non-Relativistic Limit*, Phys. Rev., 78, p. 1.
33. Feynman R.P. (1948) *Space-time Approach to Non Relativistic Quantum Mechanics*, Rev. Mod. Physics, 20, 2, p. 367.
34. Schrödinger, E. (1931) *Zur Quantendynamik des Elektrons*, Berliner, p. 63-72.
35. Schrödinger, E. (1935) *Discussion of Probability Relations between Separated Systems*, Proc. Cambridge Phil. Soc. 31, p. 555.
36. Joos. E, Zeh H.D., Kiefer C., Giulini D., Kapsh J., Stamatescu I.O. (2003) *Decoherence and the Appearance of a Classical World in Quantum Theory* (Springer).
37. Belnap N. (1992) *Branching Space-Time*, Synthese, 92, p. 385.
38. Eddington A. (1939) *Philosophy of Physical Science* (Cambridge).
39. Einstein A., Podolsky B., Rosen K. (1935) *Can Quantum-Mechanical Description of Physical Reality be Considered Complete*, Phys. Rev. 47, p. 777.
40. Grasshof G., Portmann S. and Wuthrich A. (2005) *Minimal Assumption Derivation of a Bell-Type Inequality*, Brit. J. Phil. Sci., 56, p. 663.
41. Broglie, L. (1956) *Tentative d'Interprétation Causale et Non-linéaire de la Méchanique Ondulatoire* (Gauthier-Villars).
42. Klein O. (1926) *Quantentheorie und Fünfdimensionale Relätivitatstheorie*, Zeitschrift f. Phys., 37, p. 895.
43. Sorkin R. D. (1997) *Forks in the Road, on the way to Quantum Gravity*, Int. J. Theor. Phys., 36, p. 2759.
44. Wesson P. S. (2007) *Space-Time-Matter* (World Scientific).

Index

EmerQuM 11: Emergent Quantum Mechanics 2011   IOP Publishing
Journal of Physics: Conference Series **361** (2012) 012009   doi:10.1088/1742-6596/361/1/012009

A causal net approach to relativistic quantum mechanics

R D Bateson [1]

London Centre of Nanotechnology, University College London, 17-19 Gordon St., London, WC1H 0AH, UK.

E-mail: richarddbateson@gmail.com

Abstract.
    In this paper we discuss a causal network approach to describing relativistic quantum mechanics. Each vertex on the causal net represents a possible point event or particle observation. By constructing the simplest causal net based on Reichenbach-like conjunctive forks in proper time we can exactly derive the 1+1 dimension Dirac equation for a relativistic fermion and correctly model quantum mechanical statistics. Symmetries of the net provide various quantum mechanical effects such as quantum uncertainty and wavefunction, phase, spin, negative energy states and the effect of a potential. The causal net can be embedded in 3+1 dimensions and is consistent with the conventional Dirac equation. In the low velocity limit the causal net approximates to the Schrödinger equation and Pauli equation for an electromagnetic field. Extending to different momentum states the net is compatible with the Feynman path integral approach to quantum mechanics that allows calculation of well known quantum phenomena such as diffraction.

1. Introduction

Causality, or the concept of "cause and effect", has long been viewed by philosophers as a fundamental and often an *a priori* principle in our understanding and interpretation of nature. Although Newton's laws were clearly written in causal terms, the indeterminacy of quantum mechanics lead to confusion of the role of causality in describing quantum systems. Several attempts have been made to fuse relativity and quantum indeterminism, for example, where branching space-time can be built on the primitives of a set of "possible point events" and causal relations [1] and the recent causal set approach to quantum gravity [2]. The fundamental equation of relativistic quantum mechanics is the Dirac equation for a fermion [3] and is based on the concept of continuous space and time. Richard Feynman [4] presented a discrete space-time derivation of the 1+1 dimension Dirac equation for a free particle – the "Feynman chequerboard" – since a luminal velocity massive particle is viewed in the calculation as "zig-zagging" diagonally forwards through space-time in a similar manner to a bishop in chess. Numerous attempts have been made to achieve a discrete quantum mechanics [5,6,7,8,9] but an exact lattice based formulation has never been achieved. In this paper we discuss a causal network discretisation approach which exactly derives the full 4-vector Dirac equation and provides all the common fermion features, such as spin, negative energy states, action of a potential and summation of

---

[1] Present address: Oxford Man Institute, University of Oxford, Eagle House, Walton Well Road, OX2 6ED, UK.

paths. The most basic causal net describes a plane wave solution with the space axis aligned along the direction of momentum. This 1+1 dimension net can be embedded in 3+1 dimension space-time using the Pauli matrices and is consistent with the full Dirac equation and quantum mechanical statistics.

2. Reichenbach's principle of common cause and causal networks

The application of probability theory to causality and its relation to the direction of time was developed by Hans Reichenbach. His *principle of common cause* (PCC) [10] was summarized as follows: "If coincidences of the two events A and B occur more frequently than would correspond to their independent occurrence, that is, if these events satisfy $P(A.B) > P(A)P(B)$ then there exists a common cause C for these events that the fork ACB is conjunctive." That is the probability of A and B occurring together is greater than the product of the individual probabilities of A and B. A conjunctive fork ACB (see Fig. 1) between events is open on one side where C is earlier in time than A or B. This asymmetry Reichenbach argued provides a definition of the flow of the direction of time in terms of microstatistics. Essentially a common cause is expected when coincidences or correlations between events occur repeatedly with greater frequency than complete statistical independence $P(A.B) = P(A)P(B)$. The principle of common cause provides a definition of simultaneity, since if A and B are simultaneous there cannot be a causal linkage between them except through the earlier event C. Reichenbach's principle of common cause can be readily extended into a relativistic framework [11] and a standard simultaneity condition developed [12] where simultaneous events lie on a hyperplane orthogonal to the particle world-line. It is widely accepted that quantum mechanics cannot be developed from a basic application of Reichenbach's principle of common cause with a single conjunctive fork since general quantum mechanical statistics violate the principle. The principle of common cause involving a single conjunctive fork can even actually be used to formulate the well known Bell's theorem [13,14], which motivated the famous experiments by Alain Aspect [15] to exclude the possibilities of certain types of hidden variables. Here we shall consider a modified framework where a complete causal network of possible events is comprised of conjunctive forks such that each possible event has two effective local common causes or screening factors. It appears that, at least in the simple case we consider, this allows our common cause principle - based on the simultaneity of neighbouring possible events - to be applied consistently with quantum statistics.

We shall adopt a relational view of time as an ordered series of closely spaced "events". Now if we consider time as a series of closely spaced events then from this perspective a classical particle trajectory could appear as a statistically correlated series of events in space-time (for example, a series of actual observations). If the correlation is perfect then one may loosely say that an event at one point in space "causes" the event at the next point, providing a Newtonian trajectory. However, if the correlation of events is imperfect, but greater than that resulting from statistical independence, then adjacent events in space are implied to have a common cause originating at a previous time. A trajectory becomes probabilistic in nature and we would have to involve a statistical interpretation. A network of Reichenbach's conjunctive forks [10] constitutes a causal net in which time is ordered and events may be considered simultaneous only when they share a common cause. To construct the causal net for a particle motion in space-time, we consider a 1 dimensional space aligned with the direction of particle motion, and embedded in 3 dimensional space. In this 1 dimensional space the simplest causal net that satisfies our definition of simultaneity is a 1+1 dimensional "diamond" lattice with causal links connecting the lattice points as in Figure 2. Each causal connection is defined by a connecting arrow giving a definite lineal order and an associated probability. Each vertex on the causal net represents a possible event - meaning a possible observation of the particle - and has two incoming and two outgoing causal connections so that each event has two effective possible common causes.

Starting at a vertex and following an outgoing arrow at random at each subsequent vertex describes a "causal chain" as a series of possible events. Measurement or observation at a vertex or a region of the net provides, through Bayesian statistics, a re-evaluation of these probabilities after a measurement. This is illustrated in Figure 2 where an event at A is more likely to have been caused by an event at B than C and D is an impossibility due to zero connectivity between the paths. Bayes theorem provides a way of translating this common sense concept into a formal probabilistic context since $P(A \mid B) > P(A \mid C) > P(A \mid D)$.

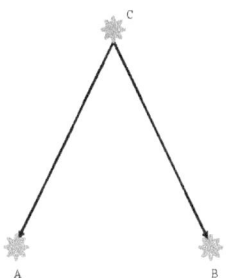

Figure 1. Reichenbach conjunctive fork linking events A and B with common cause C. C is earlier in time than simultaneous events A and B.

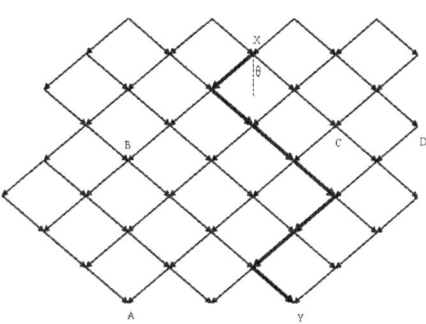

Figure 2. Causal net showing causal chain from X to Y.

3. Relativistic causal nets for a free particle

First, we will consider the simple case of a particle randomly diffusing on the causal net shown in Figure 2. In this model time and space can be discretely "counted" by attaching an integer to each of the vertex points but there is no underlying continuous space-time. To relate to conventional mechanics we interpolate this set of integers by a set of real number coordinates. Expecting that space and time have different dimensions we need to introduce a constant c with dimensions [space/time]. The net is then made up of elementary triangles labelled with $(\Delta x, c\Delta t, c\Delta \tau)$ as shown in Figure 3. We have not yet added any specific interpretation to these quantities. However, to guarantee invariance of causality on the net we impose c as the speed of light [16]. Since, from geometry, $\Delta x/c\Delta t = \sin\theta \leq 1$, we then identify Δx and Δt as relativistic space-time intervals in an observer frame S' and $\Delta \tau$ as the particle proper time interval in its rest frame S. The net geometry guarantees the invariant space-time interval

$$(c\Delta\tau)^2 = (c\Delta t)^2 - (\Delta x)^2. \tag{1}$$

Having abandoned the concept of absolute and continuous space-time we need to define the observed velocity in terms of finite differences. The definition we shall adopt is $v = \Delta x/\Delta t$ which we equate to the expectation of the velocity on the causal net. The two time intervals are then related by $\Delta \tau = \Delta t/\gamma$ where $\gamma = 1/\sqrt{1 - v^2/c^2}$ is the Lorentz factor specifying the net angles $\cos\theta = 1/\gamma$ and $\sin\theta = v/c$.

We now specialise to the case of the motion of a free particle. Clearly Eq. (1) and thus the net can be scaled by a factor. If we identify this with the particle rest mass m then from Eq. (1) we then have the relativistic dispersion relation $E^2 = p^2c^2 + m^2c^4$ where E is the particle energy $E = \gamma mc^2$ and $p = \gamma mv$ the momentum. We can further rearrange to derive a second useful invariant relation $-mc^2\Delta\tau = p\Delta x - E\Delta t$ and a third, the Lagrangian for a free particle $L = -mc^2/\gamma = pv - E = pv - H$ where H is the Hamiltonian.

From our definition of simultaneity and the geometry of the net we can see that the invariant relations provide an action $\sum p\Delta x$ which is the same on the lattice for all paths between two events. This is a restatement of Maupertuis principle, which is a weak form of the well known principle of least action, that is the integral $\int p\,dx$ is stationary. On our net if the action $\sum p\Delta x$ differed for different trajectories then this would rule some trajectories as physically inadmissible. Therefore we conclude that $\sum p\Delta x$ is the same on the lattice for all paths between two events which implies that $p\Delta x$ is a constant η for a valid causal net.

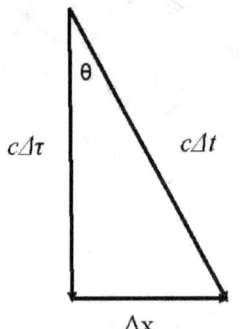

Figure 3. The elementary space-time triangle for the causal net.

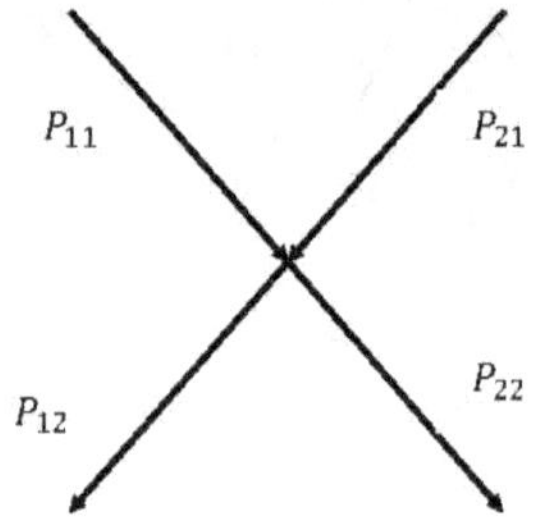

Figure 4. A vertex (1,2) on the causal net with associated probabilities.

To impose our imperfect correlation of events we shall assume that there is an indeterminism or randomness to the particle motion at each net vertex. We shall make the assumption that this indeterminism is governed by Eq. (1) on the causal net. Thus a particle in its own rest frame S over interval $\Delta\tau$ moving at a speed $|v|$ in inertial frame S' can move to a position $\pm\Delta x$ in time Δt. This produces a random trajectory in space-time (Fig. 2).

Initially we shall consider "classical" or non–quantum probabilities in construction of the causal net. Consider an individual vertex on the net and label the incoming probabilities on row 1 P_{11} and P_{21} and outgoing probabilities on row 2 P_{12} and P_{22} (Fig. 4). Probability is conserved at the vertex and the total probability at a vertex is given by $\Omega_{12} = P_{11} + P_{21}$. If the average velocity measured on the lattice is uniform then $P_{11} = P_{22}$ and $P_{12} = P_{21}$. This implies that the probabilities "cross" at each vertex without actually interfering although the probabilities are coupled. We shall see that this non-interacting case corresponds to the equilibrium case of a free particle. If we consider normalised branching probabilities at the vertex defined as $\hat{P}_{11} + \hat{P}_{21} = 1$ then since expected velocity at the vertex is defined to be v we have

$$E[v] = \gamma\frac{\Delta x}{\Delta t}[\hat{P}_{11} - \hat{P}_{21}] = v. \tag{2}$$

The branching probabilities are then given by

$$\hat{P}_{11} = \frac{E + mc^2}{2E} \qquad \hat{P}_{21} = \frac{E - mc^2}{2E}. \tag{3}$$

From this we can see that in the low velocity limit $|v| \to 0$ then $\hat{P}_{11} \to 1$ and $\hat{P}_{21} \to 0$ and in the high velocity limit $|v| \to c$ then $\hat{P}_{11} \to \hat{P}_{21} \to 1/2$. The branching ratio Γ can be written as a function of γ or the net angle θ.

$$\Gamma = \frac{\hat{P}_{11}}{\hat{P}_{21}} = \frac{E + mc^2}{E - mc^2} = \frac{\gamma + 1}{\gamma - 1} = \frac{1 + \cos\theta}{1 - \cos\theta}. \tag{4}$$

Using the branching probabilities (Eq. 3) we can write a non trivial matrix equation linking the probabilities

$$\vec{P} = \begin{pmatrix} \hat{P}_{22} \\ \hat{P}_{12} \end{pmatrix} = \begin{pmatrix} \hat{P}_{11} \\ \hat{P}_{21} \end{pmatrix} = \begin{pmatrix} 0 & \Gamma \\ 1/\Gamma & 0 \end{pmatrix} \begin{pmatrix} \hat{P}_{11} \\ \hat{P}_{21} \end{pmatrix}. \tag{5}$$

4. Relativistic quantum mechanics on the causal net

We shall now see how the causal net is compatible with the quantum mechanics of the Dirac equation for a free particle. We notice that identifying the net constant η with Planck's constant h provides the de Broglie relation $\lambda p = h$ [17] with $\Delta x = \lambda/2$, and a Heisenberg like relation $\Delta p \Delta x \sim h/2$ [18,19]. The discrete nature of the net automatically entails a de Broglie relation and an uncertainty principle. The probability is invariant in the rest frame S of the particle and we can write each probability by combining complex probability amplitudes $P_{ij} = \phi_{ij}.\phi_{ij}^*$ with

$$\phi_{ij} = \sqrt{P_{ij}}e^{-imc^2\tau/\hbar} = \sqrt{P_{ij}}e^{i(px - Et)/\hbar}, \tag{6}$$

which depends on the proper time τ at the net vertices and $x = x_{ij}$ and $t = t_{ij}$ are defined at the discrete net vertices. The phase is independent of position x for a particular τ with equivalent phase thus defining simultaneity on the net and is equivalent to a $U(1)$ global gauge invariance. We can rewrite Eq. (5) as

$$\Phi = \begin{pmatrix} \phi_{22} \\ \phi_{12} \end{pmatrix} = \begin{pmatrix} \phi_{11} \\ \phi_{21} \end{pmatrix} = \begin{pmatrix} 0 & \sqrt{\Gamma} \\ 1/\sqrt{\Gamma} & 0 \end{pmatrix} \begin{pmatrix} \phi_{11} \\ \phi_{21} \end{pmatrix}, \tag{7}$$

which can be alternatively expressed for Eq. (7) in terms of a unique transfer matrix M

$$\Phi = \begin{pmatrix} \phi_{22} \\ \phi_{12} \end{pmatrix} = \begin{pmatrix} \phi_{11} \\ \phi_{21} \end{pmatrix} = M \begin{pmatrix} \phi_{11} \\ \phi_{21} \end{pmatrix}, \tag{8}$$

defined as

$$M = \begin{pmatrix} \cos\theta & \sin\theta \\ \sin\theta & -\cos\theta \end{pmatrix} = \begin{pmatrix} 1/\gamma & v/c \\ v/c & -1/\gamma \end{pmatrix} = \frac{1}{E}\begin{pmatrix} mc^2 & pc \\ pc & -mc^2 \end{pmatrix} = \frac{H_D}{E}. \tag{9}$$

Here we recognise H_D as the Dirac Hamiltonian for a free particle [3, 20] with defined momentum p. To connect with the complete quantum mechanics we note that Eq. (9) can be put in the conventional form [21] by assuming that space-time is locally differentiable at the vertex, allowing us to use the usual momentum operator $\hat{p}$ to replace the momentum eigenvalues p. This assumption of differentiability is satisfied if we consider space-time to be a continuum of discrete causal nets, all infinitesimally displaced in space and proper time. We can write

$$\begin{pmatrix} mc^2 & c\hat{p} \\ c\hat{p} & -mc^2 \end{pmatrix} \Psi = E\Psi = i\hbar\frac{\partial\Psi}{\partial t}, \tag{10}$$

where we have replaced the probability amplitudes Φ with the familiar 2 component Dirac spinor Ψ for the free particle [21]

$$\Psi(x,t) = \begin{pmatrix} \psi_1 \\ \psi_2 \end{pmatrix} = \Phi(x,t) = A\begin{pmatrix} 1 \\ \frac{pc}{E + mc^2} \end{pmatrix} e^{i(px - Et)/\hbar}, \tag{11}$$

and A is an appropriate normalisation constant.

5. Causal net quantum symmetries, spin and Lorentz invariance

Note that the unique matrix M above is a unitary, orthogonal matrix which provides an SU(2) group transformation corresponding to an improper rotation – that is a rotation $R(\theta)$ followed by an inversion β so $M = \beta R(\theta)$. The matrix provides the transformations for the probabilities $\vec{P} = M^2\vec{P} = I\vec{P}$ and probability amplitudes $\Psi = M\Psi$. Importantly because $M^3 = M$ there exists only two levels of symmetry at the vertex and the causal net provides simultaneously both the probabilities and the underlying probability amplitudes. Since it is an improper rotation the symmetry determines a preferred axis which provides helicity along the axis of movement. If we revisit Eq. (7) and consider both possible positive and negative roots we can see that even and odd solutions with helicity $\lambda = \pm 1$ and positive energy $\epsilon = +1$ are given by

$$\Phi_{\substack{\epsilon=+1\\\lambda=+1}} = \begin{pmatrix}\sqrt{P_{11}}\\\sqrt{P_{21}}\end{pmatrix} \qquad \Phi_{\substack{\epsilon=+1\\\lambda=-1}} = \begin{pmatrix}\sqrt{P_{11}}\\-\sqrt{P_{21}}\end{pmatrix}, \tag{12}$$

corresponding to transfer matrices $M_{\epsilon=+1,\lambda=\pm1} = \beta R(\pm\theta)$. Note that in the above and following discussion we have omitted the normalisation constant and phase factor $e^{-imc^2\tau/\hbar}$ since this cancels in both sides of Eq. (8). Until now we have considered only the positive energy states, but negative energy solutions arise from the negative solution of the relativistic dispersion relation $E = \epsilon\sqrt{p^2c^2 + m^2c^4} = \epsilon|E|$ with $\epsilon = \pm 1$. This results in a reversal of the branching probabilities in Eq. (3) and two additional possible even and odd spinor solutions

$$\Phi_{\substack{\epsilon=-1\\\lambda=+1}} = \begin{pmatrix}-\sqrt{P_{21}}\\\sqrt{P_{11}}\end{pmatrix} \qquad \Phi_{\substack{\epsilon=-1\\\lambda=-1}} = \begin{pmatrix}\sqrt{P_{21}}\\\sqrt{P_{11}}\end{pmatrix}, \tag{13}$$

for transfer matrices $M_{\epsilon=-1,\lambda=\pm1} = -\beta R(\pm\theta)$. If we include the negative energy states then, by combining all 4 net solutions above (Eq. 12 and Eq. 13), we can write 4 orthogonal 4-vectors which for helicity $\lambda = \pm 1$ and energy $E = \epsilon|E|$ with $\epsilon = \pm 1$ are

$$\Psi_{p,\epsilon,\lambda=+1} = A\begin{pmatrix}1\\0\\\frac{cp}{E+mc^2}\\0\end{pmatrix}e^{i(px-Et)/\hbar} \qquad \Psi_{p,\epsilon,\lambda=-1} = A\begin{pmatrix}0\\1\\0\\\frac{-cp}{E+mc^2}\end{pmatrix}e^{i(px-Et)/\hbar}. \tag{14}$$

Using the 4 possible transfer matrices $M_{\epsilon=\pm1,\lambda=\pm1}$ then the 1+1 dimension 4-matrix Dirac equation is

$$\begin{pmatrix}mc^2 & c\beta\hat{p}\\c\beta\hat{p} & -mc^2\end{pmatrix}\Psi = E\Psi = i\hbar\frac{\partial\Psi}{\partial t}. \tag{15}$$

Now this Dirac equation and the spinor wavefunction Eq. (15) correspond to exactly the conventional 3+1 dimension Dirac spinor for the special case of the particle moving along the x-axis and with a well defined spin aligned parallel and antiparallel with the x-axis [21].

The causal net is also consistent with the extraordinarily simple Foldy-Wouthuysen representation [22] of the Dirac equation where the positive and negative energy states are decoupled through a rotation of θ (the net angle) of the Dirac Hamiltonian. For example, one Foldy-Wouthuysen 1+1 dimension state is given by the rotation through $\theta/2$ of the Dirac state Eq. (12) since $\cos(\theta/2) = \sqrt{P_{11}}$ and $\sin(\theta/2) = \sqrt{P_{21}}$

$$\Phi_{\substack{FW\\\epsilon=+1}} = R(\theta/2)\Phi_{\substack{Dirac\\\epsilon=+1\\\lambda=+1}} = \begin{pmatrix}\sqrt{P_{11}} & \sqrt{P_{21}}\\-\sqrt{P_{21}} & \sqrt{P_{11}}\end{pmatrix}\begin{pmatrix}\sqrt{P_{11}}\\\sqrt{P_{21}}\end{pmatrix} = \begin{pmatrix}1\\0\end{pmatrix}. \tag{16}$$

Importantly, in this representation, establishing an exact particle position is impossible (there is only a mean position operator) and a particle is viewed as spread out over a finite region of about a wavelength which is consistent with our causal net picture.

The causal net can also be constructed by a Lorentz boost of the Foldy-Wouthuysen states. Setting $tanh\omega = v/c = sin\theta$ then the Dirac spinor Lorentz operator $\hat{S}_L(\omega)$ acts for example as

$$\Phi_{\substack{Dirac\\\epsilon=+1\\\lambda=+1}} = \hat{S}_L\Phi_{\substack{FW\\\epsilon=+1}} = \begin{pmatrix} \cosh\omega/2 & -\sinh\omega/2 \\ -\sinh\omega/2 & \cosh\omega/2 \end{pmatrix} \Phi_{\substack{FW\\\epsilon=+1}} = \gamma^{1/2}\begin{pmatrix} \sqrt{P_{11}} & \sqrt{P_{21}} \\ \sqrt{P_{21}} & \sqrt{P_{11}} \end{pmatrix}\begin{pmatrix} 1 \\ 0 \end{pmatrix}. \quad (17)$$

Applying successive Lorentz boosts reconstructs a symmetric causal network of connected events in any reference frame so, for example $\Phi'(x',t') = \hat{S}_L(\hat{\omega})\Phi(x,t)$, for boost $\hat{\omega}$ with $\omega' = \omega + \hat{\omega}$. The Lorentz boost to any frame preserves the conservation and composition of probabilities $P_{11} + P_{21} = \gamma(P_{11} - P_{21}) = 1$ as Eq. (2) to satisfy simultaneity on the net.

6. The 3+1 dimension Dirac equation

To extend to the general 3+1 dimension case we must consider transformations of the causal net that leave it invariant under spatial direction of velocity $\vec{v}$. Using polar coordinates then for momentum $\vec{p} = |\vec{p}|(sin\vartheta cos\varphi, sin\vartheta sin\varphi, cos\vartheta)$ we can expect that the wavefunction components become dependent on the coordinates (ϑ, φ) so $\sqrt{P_{ij}}$ becomes $\sqrt{P_{ij}}\chi(\vartheta, \varphi)$. Following Dirac's convention [3] we can replace the 1 dimension momentum operator $\hat{p}$ with the 3 dimensional momentum operator $(\vec{\sigma}.\vec{p})$, formed from Pauli matrices σ_k ($k = 1, 2, 3$). By definition this momentum operator provides the relation $(\vec{\sigma}.\vec{p})\chi_\pm = |\vec{p}|\chi_\pm$ with two eigenvectors $\chi_+ = (\cos\vartheta/2, e^{i\varphi}\sin\vartheta/2)$, $\chi_- = (-e^{-i\varphi}\sin\vartheta/2, \cos\vartheta/2)$. The general solutions for the wavefunction then become from (Eq. 12 and Eq. 13) four 4-component orthogonal vectors corresponding to up and down spin $S = \pm 1/2$ with positive and negative energies $\epsilon = \pm 1$. Omitting the phase factors and normalisation constant these are

$$\Psi_{\substack{\epsilon=+1\\S=+1/2}} = \begin{pmatrix} \sqrt{P_{11}}\chi_+ \\ \sqrt{P_{21}}\chi_+ \end{pmatrix} \qquad \Psi_{\substack{\epsilon=+1\\S=-1/2}} = \begin{pmatrix} \sqrt{P_{11}}\chi_- \\ -\sqrt{P_{21}}\chi_- \end{pmatrix}, \quad (18)$$

$$\Psi_{\substack{\epsilon=-1\\S=+1/2}} = \begin{pmatrix} -\sqrt{P_{21}}\chi_+ \\ \sqrt{P_{11}}\chi_+ \end{pmatrix} \qquad \Psi_{\substack{\epsilon=-1\\S=-1/2}} = \begin{pmatrix} \sqrt{P_{21}}\chi_- \\ \sqrt{P_{11}}\chi_- \end{pmatrix}. \quad (19)$$

These are the general solutions to the conventional 3+1 dimension Dirac equation

$$\begin{pmatrix} mc^2 & c(\vec{\sigma}.\vec{p}) \\ c(\vec{\sigma}.\vec{p}) & -mc^2 \end{pmatrix}\Psi = E\Psi = i\hbar\frac{\partial\Psi}{\partial t}. \quad (20)$$

Thus for a 3 dimensional space we require all 3 Pauli matrices to construct the vector $\vec{\sigma}$ and the dimensionality of space defines the Pauli matrices. A plane wave solution to the Dirac equation has a unique velocity or space direction and the causal net is constructed along this direction in 3 dimensional space.

7. The effect of a potential on a causal net and the Pauli equation

The case we have examined is that of a free particle but we could include a potential V on the causal net since E can be replaced by $E - V$ in the construction of the lattice and the branching ratios. Returning to the 1+1 dimension case, between two media with different scalar potentials the net is compressed or stretched in space in the potential region with a form similar to Snell's law $cos\theta_2/cos\theta_1 = E/(E - V)$. We can write Eq. (10) conveniently as

$$M = \begin{pmatrix} \cos\theta_2 & \sin\theta_2 \\ \sin\theta_2 & -\cos\theta_2 \end{pmatrix} = \frac{1}{E - V}\begin{pmatrix} mc^2 & pc \\ pc & -mc^2 \end{pmatrix} = \frac{H_D}{E - V}. \quad (21)$$

If we consider invariance under a local gauge transformation U then in general $M(U\Phi) \neq (U\Phi)$ so to retain invariance we must add an additional term to the causal net Dirac equation or

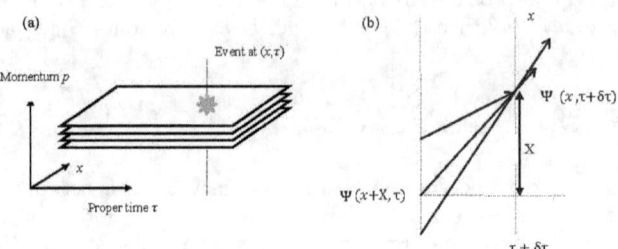

Figure 5. (a) A stacked "deck" of causal nets for different momentum states and (b) causally connected paths traversing a single space-time event.

Lagrangian which corresponds to a gauge potential term. The simple conjunctive fork (Fig. 1) is effectively replaced with an "interactive" fork [23], representing the interaction between different causal nets. To illustrate, consider the special case of a transformation where the proper time interval $\Delta\tau$ is unchanged by a potential. The triangle in Figure 3 is deformed by an amount $\delta\tau$ in time and δx in space $(c\Delta\tau)^2 = (c(\Delta t - \delta t))^2 - (\Delta x - \delta x)^2$. If we write $eA_0 = \gamma mc\delta t/\Delta t$ and $eA_1 = \gamma mc\delta x/\Delta t$ then we have the dispersion relation for an electron of charge e in an electromagnetic field (A_0, A_1) as $(E - eA_0)^2 = (pc - eA_1)^2 + m^2c^4$ and the corresponding transfer matrix M is given by

$$M = \begin{pmatrix} \cos\theta_2 & \sin\theta_2 \\ \sin\theta_2 & -\cos\theta_2 \end{pmatrix} = \frac{1}{E - eA_0} \begin{pmatrix} mc^2 & pc - eA_1 \\ pc - eA_1 & -mc^2 \end{pmatrix}. \tag{22}$$

If as Eq. (20) we embed the causal net in a continuous 3 dimensional space we can replace p with the 3 dimensional momentum operator and can consider the non-relativistic case of motion in a weak field. If we neglect the smaller component of the spinor and $E' = E - mc^2$ we have, following [20], the Pauli equation for a non-relativistic spin-1/2 particle.

$$\left(\frac{(\vec{p} - e\vec{A})^2}{2m} + eA_0 - \frac{e\hbar}{2mc}(\vec{\sigma}.\vec{H}) \right) \psi = E'\psi. \tag{23}$$

8. Momentum states and the Feynman path integral

It is interesting to consider the more general case of a range of momentum states with each momentum state occupying an individual causal net. This can be visualised in Figure 5(a) as a stacked "deck" of infinitely extended causal nets. Consider an event in at $(x, \tau + \delta\tau)$ and the prior events that are causally connected from a earlier slice of proper time at τ, which are given by different space points x from each momentum net. If we sum the different spinor components contributing to the overall probability amplitude at $(x, \tau + \delta\tau)$ and include the change in phase over interval $\delta\tau$ from Eq. (6) we have

$$\Psi(x, \tau + \delta\tau) = \sum_{\substack{causally \\ connected \\ points}} \Psi(x + X, \tau)e^{-imc^2\delta\tau/\hbar}, \tag{24}$$

where X is the relative space coordinate (Fig. 5b). For one casual net representing a free particle with a single momentum state Eq. (24) is trivial since velocity and probability are uniform across the net with only the phase varying with τ so $\Psi(x, \tau + \delta\tau) = \Psi(x + X, \tau)e^{-imc^2\delta\tau/\hbar}$

providing a simple delta function propagator for proper time interval $\delta\tau$, $K(x, x + X; \tau + \delta\tau) = \delta(X)e^{-imc^2\delta\tau/\hbar}$. However, if there is a continuum of momentum nets by geometry the sum in Eq. (24) selects a single probability amplitude contribution from each net with momentum $p = mX/\delta t$ for a given relative position X. Writing the relativistic infinitesimal action $S_{rel}(\delta\tau) = -mc^2\delta\tau$ we can write Eq. (24) as an integration

$$\Psi(x, \tau + \delta\tau) = \int_{-\infty}^{\infty} \Psi_p(x + X, \tau)e^{iS_{rel}(\delta\tau)/\hbar}\mathrm{d}X, \tag{25}$$

where Ψ_p denotes the spinor with momentum $p = mX/\delta\tau$. If we consider the non-relativistic limit where one spinor component ψ dominates and $\tau \to t$ we can use the semi-classical action $S_{cl} = \int(m/2)(\mathrm{d}x/\mathrm{d}t)^2\mathrm{d}t$ and write this infinitesimal path integral in the limit of large time interval T to give the conventional Feynman path integral [4]

$$\psi(x, t + T) = \int_{-\infty}^{\infty} K_0(x, X; T)\psi(X, t)\mathrm{d}X, \tag{26}$$

where $K_0(x, X; T) = \sqrt{(m/2\pi i\hbar T)}e^{imX^2/2\hbar T}$ is the free particle propagator for the Schrödinger equation. The causal net model is thus consistent with the quantum mechanical summation of paths and solutions to various problems such as slit diffraction using Feynman integrals [4].

9. Non-Euclidean space-time, general relativity and mass

In non-Euclidean curved space-time our elementary triangles (Fig. 3) comprising our causal net will become distorted and we can no longer apply Pythagoras' theorem to evaluate the space-time interval. In the language of general relativity the space-time interval is given by the metric $g_{\mu\nu}$ so $ds^2 = g_{\mu\nu}dx^\mu dx^\nu$. Previously, we have considered the special case of the Minkowski metric $\eta_{\mu\nu}$ for flat space-time but general relativity considers Riemann spaces that have quadratic metric equations and are characterised as locally flat. Considering a small displacement in space Δx from a point x using Taylor expansion we have the metric $g_{\mu\nu}(x + \Delta x) = \eta_{\mu\nu} + \frac{1}{2}g_{\mu\nu,\rho\sigma}\Delta x^\rho\Delta x^\sigma$. We might assume that variation to the scalar Minkowski action S_{rel} would produce a correction S_g which, to be a scalar under general coordinate transformations, can only include second order derivatives of the metric. Mathematically, the simplest curvature scalar is the Ricci scalar $R = g^{\mu\nu}R_{\mu\nu}$ formed from the Ricci curvature tensor $R_{\mu\nu}$. We can postulate that the simplest additional action might be of the form $S_g = B\int R\mathrm{d}^4x$. If we arbitrarily set the constant $B = -\varepsilon_0/16\pi Gc^4$ where ε_0 is a "density" and G Newton's gravitational constant then we have Einstein's *unimodular* gravity. If we further impose the density as $\varepsilon_0 = -\sqrt{-g}$ where $g = det(g_{\mu\nu})$ then we recover the Einstein–Hilbert action

$$S_g = -\frac{1}{16\pi Gc^4}\int R\sqrt{-g}\mathrm{d}^4x. \tag{27}$$

Thus the causal net model, although a microscopic theory, would appear to be consistent with the macroscopic theory of general relativity if the elementary triangles of our causal net model are distorted to "tile" curved space–time between causally connected possible events to preserve our definition of simultaneity.

Finally, in our model the particle mass m is an arbitrary scaling constant but if we assume a 1+1 dimension Laplace equation describes the diffusive causal paths on the net, then by imposing a Dirichlet boundary condition, the uncharged mass term of the kth fermion follows an exponential spectrum with $m_k = exp(ak + b)$ where a and b are constants.

10. Discussion

By considering simple casual connections between elementary events, based on Reichenbach's principle of common cause, we have constructed a causal model where the Dirac equation and the fermion particles it describes are seemingly "emergent" properties. The simplest causal network describes exactly the Dirac equation and provides quantum mechanical statistics and major quantum phenomena (diffraction, wave-particle duality, uncertainty ...). In an inertial frame an observer will view an ordered series of events in space–time as an entity behaving as either a wave or a particle, depending on how the experimental measuring setup is conceived, and thus exhibits wave-particle duality exactly as proposed by de Broglie. Geometrical quantities of the causal net correspond to measurable physical qualities: mass (scaling factor), momentum and energy (net angle and geometry) and potentials and forces (change of net angle). The global gauge symmetry of the net provides quantum phase and the other degenerate solutions arising from the symmetries of the net are equivalent to the Dirac spin and negative energy states. The discretisation of the net infers similarities to the Heisenberg uncertainty principle and a "stack" of nets provides, consistent with the Feynman path integral approach, quantum phenomenon such as diffraction. Distorting the causal net in non-Euclidean space-time suggests analogies with general relativity. Also imposing extra internal Lie symmetries corresponds to the case of "interacting" causal forks and leads to the introduction of gauge forces such as electromagnetism. Conventional continuous physics, with its well known differential equations, can perhaps be viewed as being "emergent" from a discrete underlying causal net. None of these emergent aspects of our causal net would have been apparent from our simple starting point of an equilibrium distribution of ordered events. Fermion particles can be considered as quasiparticles of the causal network composed of possible and actual events analogous to holes, phonons and recent magnetic monopoles in spin-ice. The quantum measurement "problem" is reduced to simple Bayesian statistics and there is no wavefunction "collapse". The quantum to classical transition becomes a straightforward feature of the resolution of the net and conventional quantum mechanics can be tentatively viewed as nature's causal "equilibrium" condition. Lastly, the model is compatible with both subjective (Kant, Hume) and objective (Reichenbach) philosophical models of causality.

References

[1] Belnap N 1992 *Synthese* **92** 385
[2] Sorkin R D 1997 *Int. J. Theor. Phys.* **36** 2759
[3] Dirac P A M. 1928 *Proc. Royal Soc.* A 117
[4] Feynman R P and Hibbs A R 1965 *Quantum Mechanics and Path Integrals* (McGraw-Hill)
[5] Kauffman L H and Noyes H P 1996 Discrete Physics and the Dirac Equation *Preprint SLAC PUB 7115*
[6] Gaveau B, Jacobson T, Kac M and Schulman L S 1984 *Phys. Rev. Lett.* **53** 419
[7] Coddens G 1995 *Found. Physics Lett.* **8** 3
[8] Bialynicki Birula I 1994 *Phys. Rev.* D **49** 12
[9] Schotte K D, Iwabuchi S and Truong T T 1985 *Z. Phys.* B **60** 255
[10] Reichenbach H 1956 *The Direction of Time* (University California Press)
[11] Penrose O and Percival I C 1962 *Proc. Phys. Soc.* **79** 605
[12] Malament D 1977 *Nous* **11** 293
[13] Bell J 1966 *Rev. Mod. Phys.* **38** 447
[14] Grasshof G, Portmann S and Wuthrich A 2005 *Brit. J. Phil. Sci.* **56** 663
[15] Aspect A, Dalibard J, Roger G 1982 *Phys. Rev. Lett.* **49** 1804
[16] Rindler W 1982 *Introduction to Special Relativity* (Oxford)
[17] De Broglie L 1925 *Ann. Phys.* **3** 22
[18] Heisenberg W 1927 *Z. Phys.* **43** 172
[19] Schurmann T and Hoffmann I 2009 *Found. Phys.* **39** 958
[20] Davydov A S 1965 *Quantum Mechanics* (Pergamon Press)
[21] Coulter B and Adler C 1971 *AJP* **39** 305
[22] Foldy L and Wouthuysen S 1950 *Phys. Rev.* **78** 1
[23] Salmon W C 1998 *Causality and Explanation* (Oxford)

A causal net approach to gravitation and dark energy

R. D. Bateson

Cavendish Laboratory, J J Thomson Avenue, Cambridge, CB3 0HE, UK.

E-mail: rb2009@cam.ac.uk

Abstract.
   In this paper we discuss a causal network model to describe gravitation, where space–time is built solely from point events and connecting probabilities. Each vertex on the causal net represents a possible point event or particle observation. Simultaneity on the causal net is defined by hyperplanes of equivalent proper time. The causal net model when constructed in Minkowksi space–time is shown to lead to a formulation for relativistic quantum mechanics including Dirac equation. In a curved Riemann space–time the causal paths of the causal net are geodesics and in the local Lorentz frame the net probabilities and the Dirac formalism are preserved. The variation of space–time density of events in the causal paths modifies the metric and provides a space–time curvature leading to the Hilbert action associated with general relativity. The causal net model can also be applied to cosmological models of the universe and leads to cosmological redshifts and universal energy densities equivalent to the critical value for flatness. Lastly, a causal net route to dark energy is proposed where a quantum hydrostatic pressure, proportional to the net reversal "Zitterbewegung" probabilities, provides a ratio of matter energy to pressure energy close to that observed experimentally. Dark matter is then considered as the combined density of unobserved events in different orthogonal spatial directions, at each net vertex.

1. Introduction

In the causal net approach to relativistic quantum mechanics [1] space–time is built on the primitives of a set of "possible point events" and causal relations. The causal net is constructed in the local Lorentz frame and simultaneity on the causal net is defined for events lying on hyperplanes of equivalent proper time. The causal net space–time discretisation method exactly derives the Dirac equation [2] and provides all the common fermion features, such as spin, negative energy states, action of a potential and summation of paths [3]. The most basic causal net describes a plane wave solution with the space axis aligned along the direction of momentum. The causal net is similar to the Lorentz invariant *Aether* of causally connected events proposed by Dirac in 1951 [4] and is consistent with Dirac's gauge which describes the existence and motion of a classical electron with no self interaction [5].

   Einstein himself speculated on the possibility of a causal theory underlying general relativity and in 1916 he wrote "Causal set theory arises by combining discreteness and causality to create a substance that can be the basis of a theory of quantum gravity. Spacetime is thereby replaced by a vast assembly of discrete "elements" organised by means of "relations" between them...None of the continuum attributes of spacetime, neither metric, topology nor differentiable structure are retained, but emerge it is hoped as approximate concepts at macroscopic scales."

   For curved space–time the causal paths of the causal net are geodesics. In the general relativity formalism the curved space is mapped by the causal net, as originally envisaged by Einstein, as a grid of geodesics. The Dirac equation in the local Lorentz frame is unchanged in these geodesic coordinates and the geometry and probabilities linking events are preserved. The curvature of space–time is produced by varying the space–time density of events leading to a modified coordinate system for observers. Overlapping events from different causal nets, can increase the density of events and provide a causal net analogue to gravitation.

The causal net model can also be applied to cosmological models of the universe such as the Friedman model [6,7], if we assume the universe is governed as a quantum state. In the causal net model the universe energy density is always the critical value for flatness since it is related to the density of the local Lorentz frame. Dark energy is potentially described by a quantum pressure from "Zitterbewegung" and is proportional to the reversal probabilities on the causal net. Applying a cosmological fluid model to causal nets an equilibrium ratio of matter energy to pressure energy is estimated to be close to that observed experimentally.

2. The causal net approach to relativistic quantum mechanics

As a preliminary I will present an overview of the causal net approach to relativistic quantum mechanics detailed in [1]. To construct the causal net for a particle motion in space–time, consider a 1 dimensional space aligned with the direction of particle motion, and embedded in 3+1 dimensional space. In this 1 dimensional space the simplest causal net that satisfies the definition of simultaneity is a 1+1 dimensional "diamond" lattice with causal links connecting the lattice points as in Figure 3. Each causal connection is defined by a connecting arrow giving a definite lineal order and an associated probability. Each vertex on the causal net represents a possible event — meaning a possible observation of the particle — and has two incoming and two outgoing causal connections so that each event has two effective possible common causes. Starting at a vertex and following an outgoing arrow at random at each subsequent vertex describes a "causal chain" as a series of possible events.

Measurement or observation at a vertex or a region of the net provides, through Bayesian statistics, a re-evaluation of these probabilities after a measurement. A causal net of possible events thus constitutes a simple Bayesian network. This is illustrated in Figure 1 where an event at $\mathcal{A}$ is more likely to have been caused by an event at $\mathcal{B}$ than $\mathcal{C}$ and $\mathcal{D}$ is an impossibility due to zero connectivity between the paths. Bayes theorem provides a way of translating this common sense concept into a formal probabilistic context since $P(\mathcal{A} \mid \mathcal{B}) > P(\mathcal{A} \mid \mathcal{C}) > P(\mathcal{A} \mid \mathcal{D})$. On measurement of an event this Bayesian re-evaluation and reassessment of probabilities is analogous to the well known "collapse of the wavefunction" in other interpretations of quantum mechanics.

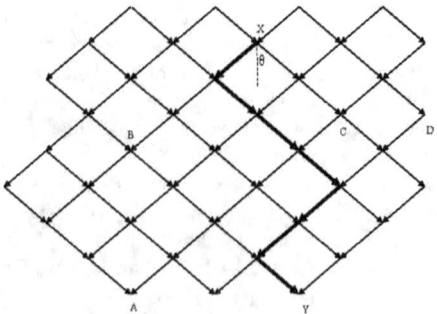

Figure 1. Causal net showing causal chain from X to Y. The net angle θ is shown.

First, consider the simple case of representing a particle randomly diffusing on the 1+1 dimension causal net shown in Figure 1. The net is then made up of elementary triangles labelled with $(\Delta x, c\Delta t, c\Delta \tau)$ as shown in Figure 2. To guarantee invariance of causality on the net we impose c as the speed of light [8]. Since, from geometry, $\Delta x/c\Delta t = \sin\theta \leq 1$, and then identify Δt as relativistic time intervals in an observer frame and $\Delta \tau$ as the particle proper time interval. The proper time interval is the actual time experienced by a particle or a local Lorentz observer moving between the two events and simultaneity on the causal net is then defined for events lying on hyperplanes of equivalent proper time . The net geometry guarantees the invariant space–time interval

$$(c\Delta\tau)^2 = (c\Delta t)^2 - (\Delta x)^2. \tag{1}$$

On the discrete causal net we define the observed velocity in terms of finite differences. The definition adopted is $v = \Delta x/\Delta t$ which we equate to the expectation of the velocity on the causal net. The two time intervals are then related by $\Delta \tau = \Delta t/\gamma$ where $\gamma = 1/\sqrt{1 - v^2/c^2}$ is the Lorentz factor specifying the net angle θ

$$\cos\theta = 1/\gamma \qquad \sin\theta = v/c. \tag{2}$$

Associated with each event the 4-vector velocity $v_\mu = \gamma(c, \vec{v})$ provides a commuting invariant relation

$$v^\mu v_\mu = c^2. \tag{3}$$

This invariant relation is the starting point for developing the mathematics of a causally connected series of events in space–time and was first proposed in 1951 by Dirac [4] as defining a Lorentz invariant Aether. In Dirac's formulation at each point in space-time the velocity of the Aether is subject to quantum uncertainty and is potentially multi–valued. The velocity is badly defined but may be described by a probability distribution and a pure isotropic empty vaccuum state is not measurable. We shall see that Dirac's Aether is consistent with the causality described by our Lorentz invariant network of possible events in space–time.

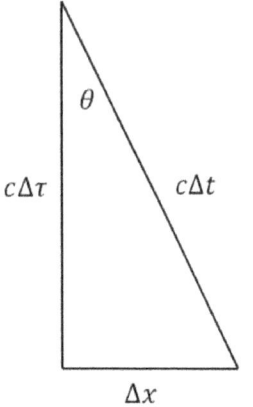

Figure 2. The elementary space-time "triangle" for the causal net built in the local Lorentz frame .

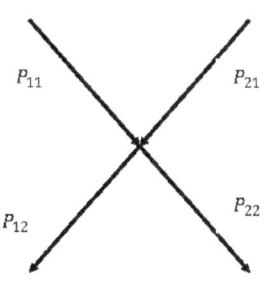

Figure 3. A vertex (1,2) on the causal net with associated probabilities.

Clearly Eq. (1) and thus the net can be scaled by a factor. If we identify this with the particle rest mass m then from Eq. (1) we then have the relativistic dispersion relation

$$E^2 = p^2 c^2 + m^2 c^4, \tag{4}$$

where E is the particle energy $E = \gamma m c^2$ and $p = \gamma m v$ the momentum. We can further rearrange to derive a second useful invariant relation

$$-mc^2 \Delta\tau = p\Delta x - E\Delta t, \tag{5}$$

and a third, the Lagrangian for a free particle

$$L = -mc^2/\gamma = pv - E = pv - H, \tag{6}$$

where H is the Hamiltonian.

If we identify the causal net as representing an electron with charge e and velocity v moving in an electromagnetic potential (A_0, A_1) with

$$eA_0 = E, \tag{7}$$

and

$$eA_1 = pc. \tag{8}$$

The net automatically describes the movement of an electron in an electromagnetic potential with the dispersion relation

$$-m^2 c^4 = e^2 A_1{}^2 - e^2 A_0{}^2. \tag{9}$$

The above 1+1 dimension case is for the field component A_1 aligned along the causal net x space axis. In 3+1 dimensions this corresponds to Dirac's gauge condition [5]

$$A^\mu A_\mu = k^2 = (mc^2/e)^2, \tag{10}$$

where the electron velocity at different points in space–time is directly linked to the local electromagnetic potential. In Dirac's classical theory there is no electron without momentum and without a field and vice versa. The gauge connects directly with Maxwell's equations for the electromagnetic field.

From our definition of simultaneity and the geometry of the net the $\sum p\Delta x$ is the same on the lattice for all paths between two events which implies that $p\Delta x$ is a constant for a valid causal net. Identifying the net constant with Planck's constant h provides the de Broglie relation [9] with $\Delta x = \lambda/2$

$$\lambda p = h, \tag{11}$$

and a Heisenberg like relation [10,11]

$$\Delta p \Delta x \sim h/2. \tag{12}$$

The discrete nature of the net automatically entails a de Broglie relation and an uncertainty principle. The spatial separation of events for a given proper time is equal to the de Broglie wavelength.

Starting with the relation Eq. (5) we can consider the causal path sum of this relation over n events between two non adjacent distant event which leads to several action principles. The Maupertuis action S_M, that does not explicitly involve the time taken between events in space, can be written as a function of the number of steps between events n and h

$$S_M = \sum_i^n p\Delta x = nh. \tag{13}$$

We can also derive a more general action principle S_τ based on proper time and related to the Lagrangian

$$S_\tau = -\sum_i^n mc^2 \Delta \tau = -mc^2 n \Delta \tau = -\sum_i^n L\Delta t, \tag{14}$$

which can approximated for large n with an integral to acquire the well known result

$$S = -\int L\mathrm{d}t = -\int mc^2 \mathrm{d}\tau. \tag{15}$$

The indeterminism on the causal net is governed by Eq. (1). Thus a particle in its own local Lorentz frame in a proper time interval $\Delta \tau$ moving at a speed $|v|$ can move to a position $\pm \Delta x$ over time Δt in an inertial observer frame. This produces a random trajectory in space–time as in Figure 1. Consider an individual vertex on the net and label the incoming probabilities on row 1 P_{11} and P_{21} and outgoing probabilities on row 2 P_{12} and P_{22} as shown in Figure 3. Probability is conserved at the vertex and the total probability at a vertex is given by $P_T = P_{11} + P_{21}$. If the average velocity measured on the lattice is uniform then $P_{11} = P_{22}$ and $P_{12} = P_{21}$. If we consider normalised branching probabilities at the vertex defined as $\hat{P}_{11} + \hat{P}_{21} = 1$ then since expected velocity at the vertex is defined to be v we have

$$E[v] = \gamma \frac{\Delta x}{\Delta t}[\hat{P}_{11} - \hat{P}_{21}] = v. \tag{16}$$

The branching probabilities are then given by

$$\hat{P}_{11} = \frac{E + mc^2}{2E} \qquad \hat{P}_{21} = \frac{E - mc^2}{2E}. \tag{17}$$

From this we can see that in the low velocity limit $|v| \to 0$ then $\hat{P}_{11} \to 1$ and $\hat{P}_{21} \to 0$ and in the high velocity limit $|v| \to c$ then $\hat{P}_{11} \to \hat{P}_{21} \to 1/2$. At higher velocities the causal path followed becomes more random and the trajectory exhibits "Zitterbewegung". The expected velocity v is equivalent in both local Lorentz and observer frames if time and space are represented as orthogonal basis vectors. The probabilities can also be written simply in terms of net angles

$$\hat{P}_{11} = \cos^2(\theta/2) \qquad \hat{P}_{21} = \sin^2(\theta/2). \tag{18}$$

Each real probability can be formed by combining complex probability amplitudes $P_{ij} = \phi_{ij}.\phi_{ij}^*$ with

$$\phi_{ij} = \sqrt{P_{ij}} e^{-imc^2 \tau/\hbar} = \sqrt{P_{ij}} e^{i(px - Et)/\hbar}, \tag{19}$$

which depend on the proper time τ at the net vertices and x and t are defined at the discrete net vertices. The probability amplitudes at each vertex on the net can be expressed in terms of a unique transfer matrix M

$$\Phi = \begin{pmatrix} \phi_{22} \\ \phi_{12} \end{pmatrix} = \begin{pmatrix} \phi_{11} \\ \phi_{21} \end{pmatrix} = M \begin{pmatrix} \phi_{11} \\ \phi_{21} \end{pmatrix}, \tag{20}$$

defined as

$$M = \begin{pmatrix} \cos\theta & \sin\theta \\ \sin\theta & -\cos\theta \end{pmatrix} = \begin{pmatrix} 1/\gamma & v/c \\ v/c & -1/\gamma \end{pmatrix} = \frac{1}{E}\begin{pmatrix} mc^2 & pc \\ pc & -mc^2 \end{pmatrix} = \frac{H_D}{E}. \tag{21}$$

Here we recognise H_D as the Dirac Hamiltonian for a free particle [4, 12] with defined momentum p. To connect with the complete quantum mechanics we note that Eq. (21) can be put in the conventional form [13] by assuming that space–time is locally differentiable at the vertex, allowing us to use the usual momentum operator $\hat{p}$ to replace the momentum eigenvalues p. We can write

$$\begin{pmatrix} mc^2 & c\hat{p} \\ c\hat{p} & -mc^2 \end{pmatrix} \Psi = E\Psi = i\hbar \frac{\partial \Psi}{\partial t}, \tag{22}$$

where we have replaced the probability amplitudes Φ with the familiar 2 component Dirac spinor Ψ for the free particle [12]

To extend to the general 3+1 dimension case we must consider transformations of the causal net that leave it invariant under spatial direction of velocity $\vec{v}$. Using polar coordinates then for momentum $\vec{p} = |\vec{p}|(sin\vartheta cos\varphi, sin\vartheta sin\varphi, cos\vartheta)$ the wavefunction components become dependent on the coordinates (ϑ, φ) so $\sqrt{P_{ij}}$ becomes $\sqrt{P_{ij}}\chi(\vartheta, \varphi)$ where $\chi(\vartheta, \varphi)$ is a multiplicative function. Following Dirac's convention [2] we can replace the 1 dimension momentum operator $\hat{p}$ with the 3 dimensional momentum operator $(\vec{\sigma}.\vec{p})$, formed from Pauli matrices σ_k $(k = 1, 2, 3)$. By definition this momentum operator provides the relation $(\vec{\sigma}.\vec{p})\chi_\pm = |\vec{p}|\chi_\pm$ with two eigenvectors

$$\chi_+ = (\cos \vartheta/2, e^{i\varphi} \sin \vartheta/2) \qquad \chi_- = (-e^{-i\varphi} \sin \vartheta/2, \cos \vartheta/2). \tag{23}$$

The general solutions for the wavefunction then become four 4-component orthogonal vectors corresponding to up and down spin $S = \pm 1/2$ with positive and negative energies $\epsilon = \pm 1$. Omitting the phase factors and normalisation constant these are

$$\Psi_{\substack{\epsilon=+1 \\ S=+1/2}} = \begin{pmatrix} \sqrt{P_{11}}\chi_+ \\ \sqrt{P_{21}}\chi_+ \end{pmatrix} \qquad \Psi_{\substack{\epsilon=+1 \\ S=-1/2}} = \begin{pmatrix} \sqrt{P_{11}}\chi_- \\ -\sqrt{P_{21}}\chi_- \end{pmatrix}, \tag{24}$$

and

$$\Psi_{\substack{\epsilon=-1 \\ S=+1/2}} = \begin{pmatrix} -\sqrt{P_{21}}\chi_+ \\ \sqrt{P_{11}}\chi_+ \end{pmatrix} \qquad \Psi_{\substack{\epsilon=-1 \\ S=-1/2}} = \begin{pmatrix} \sqrt{P_{21}}\chi_- \\ \sqrt{P_{11}}\chi_- \end{pmatrix}. \tag{25}$$

These are the common solutions to the conventional 3+1 dimension Dirac equation

$$\begin{pmatrix} mc^2 & c(\vec{\sigma}.\vec{p}) \\ c(\vec{\sigma}.\vec{p}) & -mc^2 \end{pmatrix} \Psi = E\Psi = i\hbar \frac{\partial \Psi}{\partial t}. \tag{26}$$

3. Inertial mass on the causal net

The causal net and the separation of events in space and time leads naturally to the concept of mass as being the density of possible events in space–time. Since we have fixed the net constant as Planck's constant h, the absolute size or scale of the causal net in time and space can thus be evaluated as

$$\Delta\tau = \frac{h}{2mc^2(\gamma^2 - 1)} \qquad \Delta x = \frac{h}{2mc\sqrt{\gamma^2 - 1}}, \tag{27}$$

with the scaling relation

$$\Delta\tau = \frac{2mc}{h}(\Delta x)^2. \tag{28}$$

Thus the net decreases in size with increasing velocity or energy as the de Broglie wavelength decreases. Also the scale of the net is also inversely proportional to the mass. Particles of higher mass will have more finely resolved causal nets. Importantly, objects that are large relative to the net size will transition to a more classical behaviour for observers and this resolution effect in the causal net approach provides a *correspondence principle* between the classical and quantum domains.

From Eq. (27) for the distance between events on the causal net we can see that for equivalent velocity, and hence similar net triangles, the net size scales inversely with mass. This is shown schematically in Figure 4. For two masses m_1 and m_2 we have

$$\frac{m_1}{m_2} = \frac{\Delta x_2}{\Delta x_1} = \frac{\Delta t_2}{\Delta t_1}. \tag{29}$$

The number density of events on the net with distance or time is thus proportional to the particle mass. If σ_1 and σ_2 are the density of events in space or time their ratio is given by

$$\frac{\sigma_1}{\sigma_2} = \frac{m_1}{m_2}. \tag{30}$$

Thus, in the causal net model the concept of inertial mass is thus directly equivalent to the density of possible causal events.

We can see that the concept of mass on a causal net and simultaneity provides a link to Mach's principle, whereby it was postulated distant events or far away nebula can influence physics on a local scale. From Eq. (27) for causal paths with relativistic velocities, the size of the causal net is reduced and the net angle $\theta = \arcsin(\frac{v}{c})$ increases towards $\frac{\pi}{2}$. For a free particle, causal events over a finite proper time τ but that are at very large cosmic distances (since $x = c\tau \tan\theta$) can have a causal influence on local events whilst remaining simultaneous with closer events. However, due to the smaller net size such spatially distant causal nets must traverse many intermediate events on their passage and their probability of influencing local events is much lower than for nearby events as in Figure 5.

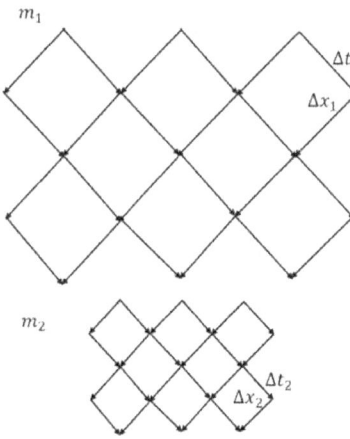

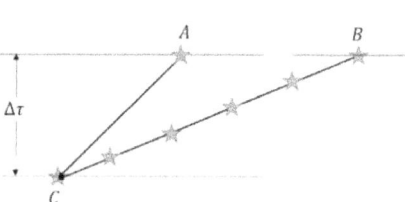

Figure 5. An event at B may be much further away in space and time than event A from event C but may be simultaneous in proper time and have an influence on C despite traversing many events in the causal path.

Figure 4. Causal nets for same velocity and hence net angle for mass m_1 and m_2 where $m_2 > m_1$.

4. Causal nets and general relativity

We shall assume that space-time is Reimann as assumed in the general relativity of Einstein. In the language of general relativity the space–time interval is given by the metric $g_{\mu\nu}$ so $ds^2 = g_{\mu\nu} dx^\mu dx^\nu$. Previously, we have considered the special case of the Minkowski metric

$\eta_{\mu\nu}$ for flat space–time but general relativity considers Riemann spaces that have quadratic metric equations and are characterised as locally flat. The causal net is built in the local Lorentz frame with geodesics connecting possible events. A hypothetical observer moving along the path between two events experiences motion along a straight line with constant velocity. We can construct a causal net for Minkowski space–time and then embed the causal net into curved space–time with new coordinates. The coordinates of events in any curved space–time will thus change for any observer not in the geodesic local Lorentz frame. The curved space–time will then be criss-crossed by a grid of geodesics linking events.

The causal net for each velocity is built using a so-called Riemann normal coordinate system [7] with the following characteristics close to an event at $\mathcal{P}$

$$g_{\alpha\beta}(\mathcal{P}) = \eta_{\alpha\beta} \tag{31}$$

$$g_{\alpha\beta,\mu}(\mathcal{P}) = 0. \tag{32}$$

Thus close to any point or event the Riemann space–time is effectively Minkowski and described by the Minkowski metric. Further relations can be derived for the case of normal coordinates close to $\mathcal{P}$ to give the Riemann tensor as the second order derivative of the metric. The causal net of events and the probabilities connecting events, and hence the spinor components will be preserved in moving from flat to curved space–time since there exists local Lorentz invariance in the vicinity of every possible event in Reimann space. This means that at each event space–time is Euclidean $g_{\alpha\beta}(\mathcal{P}) = \eta_{\alpha\beta}$ and the angle θ of the causal net is preserved for a given geodesic velocity. The path visibly taken between events in the observer frame is irrelevant for a geodesic local Lorentz observer that travels along a straight path with constant velocity and arrives at each event with the unchanged net angle θ. The transfer matrix M for the net Eq. (21) and the subsequent results extending to 3+1 dimensions remain unchanged. At each event the branching probabilities are preserved for the plane wave state and hence the Dirac equation is maintained. This representation of the Dirac equation in the local Lorentz frame is much simpler than the representation in the inertial frame of an arbitrary observer who must impose a suitable basis and coordinate system for measurement. In the latter case, mathematical techniques called verbein fields are often introduced to transform to the new frame and coordinate system and this leads to complex forms of the Dirac equation. However, the key result here is that in the geodesic frame, represented by the causal net, the Dirac equation is unchanged and retains its simple, elegant form.

As for the Minkowski metric, for simultaneity we must consider hyperplanes between events that have constant proper time, that is the time experienced travelling along the geodesic between events is the same. A path of extremal τ is straight with constant velocity in every local Lorentz frame and is a geodesic of space–time. In general, for curved space–time we can conventionally write the proper time interval between events $\mathcal{A}$ and $\mathcal{B}$ as

$$\tau = \int_{\mathcal{B}}^{\mathcal{A}} d\tau = -\int_{\mathcal{B}}^{\mathcal{A}} (g_{\mu\nu} dx^\mu dx^\nu)^{1/2}. \tag{33}$$

By standard deformation and maximising or extremal lapse of proper time [6,7] we have the geodesic equation in general coordinates at an event $\mathcal{P}$ at x^α as

$$\frac{\partial^2 x^\alpha}{\partial \tau^2} = -\Gamma^\alpha{}_{\mu\nu} \frac{\partial x^\mu}{\partial \tau} \frac{\partial x^\nu}{\partial \tau} = 0. \tag{34}$$

At the event $\mathcal{P}$ we have $-\Gamma^\alpha{}_{\mu\nu} = 0$ leading to a secondary relation

$$g_{\alpha\beta} \frac{\partial x^\alpha}{\partial \tau} \frac{\partial x^\beta}{\partial \tau} = -c^2. \tag{35}$$

From this equation we can see how the metric $g_{\alpha\beta}$ influences the coordinate system x^{α} for the causal net. To illustrate, we return to the 1+1 dimension causal net case and assume that observers can chose a orthogonal local Lorentz coordinate system to diagonalise the metric. This diagonalisation separates the time and space components of the metric. For flat space–time we have

$$-(c\Delta\tau)^2 = \eta_{00}(c\Delta t)^2 + \eta_{11}(\Delta x)^2, \tag{36}$$

and for curved space–time in 1+1 dimension with time and space coordinates $\hat{t}$ and $\hat{x}$,

$$-(c\Delta\tau)^2 = g_{00}(c\Delta\hat{t})^2 + g_{11}(\Delta\hat{x})^2. \tag{37}$$

If simultaneity between events on the net is determined, as before, by their equivalent separation in proper time $\Delta\tau$, these equations can be rearranged to give

$$\gamma^2\left(c^2 + \frac{\eta_{11}(\Delta x)^2}{\eta_{00}(\Delta t)^2}\right) = \gamma^2\left(c^2 + \frac{g_{11}(\Delta\hat{x})^2}{g_{00}(\Delta\hat{t})^2}\right) = v^{\mu}v_{\mu} = c^2. \tag{38}$$

Provided the proper time interval is preserved $\Delta\tau$ and the net velocity is equivalent to the ratio

$$\left(\frac{\eta_{11}(\Delta x)}{\eta_{00}(\Delta t)}\right) = \left(\frac{g_{11}(\Delta\hat{x})}{g_{00}(\Delta\hat{t})}\right) = -v, \tag{39}$$

the causal net paths remain geodesics of constant velocity. Any geodesic trajectory through curved space–time can be mapped piecewise to a geodesic causal net built in the local Lorentz frame with net triangles as in Figure 6.

This generalises to the previous geodesic relation Eq. (33) for the observer frame in 3+1 dimensions if time and space can be separated. The use of orthogonal local Lorentz coordinate systems allows time and space to be separated in obeying simultaneity. This results in the causal net obeying the relativistic energy dispersion relations in all measurable reference frames. Experimentally, measurements of the universality of the relativistic energy dispersion relations, such as those of electrons in the Crab Nebula [14] have placed tight constraints on any deviations from the well known energy dispersion relations, as Eq. (4), and their invariant quantities.

The above consideration shows that if a causal net is constructed using geodesics to separate neighbouring events with space–time separations Δx^{α} and if a metric is applied to curve space then new coordinates $\Delta\hat{x}^{\alpha}$ must be applied to each event on the net. This effectively "bends" the net in space–time for observers. For the 1+1 dimension case, the distance between events changes such that

$$\Delta t = \Delta\hat{t}\sqrt{g_{00}} \qquad \Delta x = \Delta\hat{x}\sqrt{g_{11}}, \tag{40}$$

and the $\Delta\hat{t}$ and $\Delta\hat{x}$ no longer have a Pythagorean relationship in constructing the causal net but require a metric scaling factor as in Figure 6.

Until now we have treated the metric as an abstract entity, however, the metric is provided by the density of events along each causal path. To see this consider the path between two events of total proper time τ for a particle of mass m_1 which comprises n_1 events in the local Lorentz frame. We can write in 1+1 dimensions for a causal net

$$-(\tau c)^2 = -(n_1 c\Delta\tau_1)^2 = \eta_{00}(n_1 c\Delta t_1)^2 + \eta_{11}(n_1\Delta x_1)^2. \tag{41}$$

We can define the metric to be $g_{001} = \eta_{00}n_1^2$ and $g_{111} = \eta_{11}n_1^2$. The absolute number of events from Eq. (22) can be written as

$$n_1 = \frac{\tau}{\Delta\tau_1} = \frac{2m_1c^2(\gamma^2 - 1)\tau}{h}. \tag{42}$$

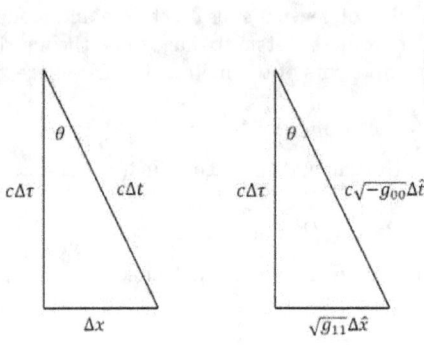

Figure 6. The causal net triangles are equivalent in the local Lorentz frame (left) and the observer frame (right) when scaled by the space–time metric. Curved space–time for geodesics can be mapped to a Minkowski causal net.

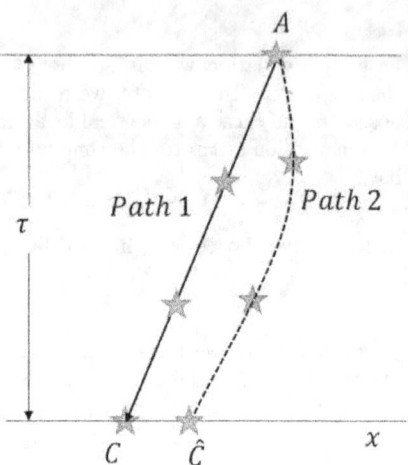

Figure 7. In curved space–time the density of events varies and this leads to an effective metric if simultaneity in proper time is preserved for different causal paths. The number of events in each path (Path 1 and Path 2) is equal for a particular particle mass but the space–time separation between them varies.

For a second particle with mass m_2 and equivalent velocity which comprises n_2 events to cover the same proper time τ

$$-(\tau c)^2 = -(n_2 c \Delta \tau_2)^2 = \eta_{00}(n_2 c \Delta t_2)^2 + \eta_{11}(n_2 \Delta x_2)^2. \tag{43}$$

The ratio of events is then from Eq. (25)

$$\frac{n_1}{n_2} = \frac{m_1}{m_2} = \frac{\sigma_1}{\sigma_2}, \tag{44}$$

where σ is the density of events and is proportional to the mass. We can write the ratio of metrics, relative masses, density of events and causal net space–time distances conveniently as

$$\frac{\sqrt{-g_{00}}_1}{\sqrt{-g_{00}}_2} = \frac{\sqrt{g_{11}}_1}{\sqrt{g_{11}}_2} = \frac{n_1}{n_2} = \frac{m_1}{m_2} = \frac{\sigma_1}{\sigma_2} = \frac{\Delta x_2}{\Delta x_1} = \frac{\Delta t_2}{\Delta t_1}. \tag{45}$$

In this special symmetric case where $g_{00} = -g_{11}$ the metric is a scaling in space–time proportional to the mass or number of events squared and the Minkowski metric is simply multiplied by a scaling factor. It is tempting to absorb the scaling factor into the Minkowski metric such that the metric components are never unity. Indeed, this was proposed by Feynman, who suggested, that Minkowski metrics might be of the form $\eta_{00} = -\xi$ and $\eta_{ii} = \xi$ where ξ is not unity [15].

For Minkowski space–time events are distributed homogeneously but the more general case might be where the density of events varies differently in time σ_t and space σ_x. The path can now be written in the new coordinates $\hat{t}$ and $\hat{x}$ as the ratio of density of events between the curved and flat space–time

$$-(\tau c)^2 = -(n_1 c \Delta \tau_1)^2 = \eta_{00}(n_1 \frac{\hat{\sigma}_{t1}}{\sigma_1} c \Delta \hat{t}_1)^2 + \eta_{11}(n_1 \frac{\hat{\sigma}_{x1}}{\sigma_1} \Delta \hat{x}_1)^2. \tag{46}$$

The metric then becomes $g_{00_1} = \eta_{00}(n_1 \frac{\hat{\sigma}_{t1}}{\sigma_1})^2$ and $g_{11_1} = \eta_{11}(n_1 \frac{\hat{\sigma}_{x1}}{\sigma_1})^2$ for 1+1 dimensions. The number of events n_1 is the same over the proper time interval τ but the distance in space and time between them scales with the density of events such that

$$\frac{\sqrt{-\hat{g}_{00_1}}}{\sqrt{-g_{00_1}}} = \frac{\Delta t_1}{\Delta \hat{t}_1} = \frac{\hat{\sigma}_{t1}}{\sigma_1} \qquad \frac{\sqrt{\hat{g}_{11_1}}}{\sqrt{g_{11_1}}} = \frac{\Delta x_1}{\Delta \hat{x}_1} = \frac{\hat{\sigma}_{x1}}{\sigma_1}. \tag{47}$$

The above results are similar to the well known continuous time definition for the metric based on differentials

$$g_{rs} = \delta_{mn} \frac{\partial x^m}{\partial \hat{x}^r} \frac{\partial x^n}{\partial \hat{x}^s}, \tag{48}$$

but the causal net result assumes a choice of inertial frame coordinates that diagonalises the metric and is based on finite differences.

This methodology extends to 3+1 dimensions and we can write the proper time between events as

$$-(\tau c)^2 = -(n_1 c \Delta \tau)^2 = \eta_{00}(n_1 c \Delta t_1)^2 + \eta_{11}(n_1 \Delta r_1)^2, \tag{49}$$

where $\Delta r_1{}^2 = \Delta x_1{}^2 + \Delta y_1{}^2 + \Delta z_1{}^2 = \sum(\Delta x_{ii_1})^2$. For the curved space–time we have

$$-(\tau c)^2 = -(n_1 c \Delta \tau)^2 = g_{00_1}(c \Delta \hat{t}_1)^2 + \sum g_{ii_1}(\Delta \hat{x}_{ii_1})^2, \tag{50}$$

where

$$\frac{\sqrt{-\hat{g}_{00_1}}}{\sqrt{-g_{00_1}}} = \frac{\Delta t_1}{\Delta \hat{t}_1} = \frac{\hat{\sigma}_{t1}}{\sigma_1} \qquad \frac{\sqrt{\hat{g}_{ii_1}}}{\sqrt{g_{ii_1}}} = \frac{\Delta x_{i_1}}{\Delta \hat{x}_{i_1}} = \frac{\hat{\sigma}_{i_1}}{\sigma_{i_1}}. \tag{51}$$

The 4-volume elements $\Delta x^4 = \Delta t \Delta x \Delta y \Delta z$ in the two frames are related as

$$\frac{\sqrt{-\hat{g}_1}}{\sqrt{-g_1}} = \frac{\Delta x_1^4}{\Delta \hat{x}_1^4}, \tag{52}$$

where g is the trace of the metric. This leads to the common connection between the proper volume in the local Lorentz frame and the observer frame coordinates $d^4 x = \sqrt{-g} d^4 \hat{x}$. The total density of events for each metric multiplied by the 4-volume element is itself constant and since the number of events is proportional to a mass energy this is a restatement of the conservation of energy density.

In most physical situations the metric changes in space and time and this is equivalent to the density of events varying. We have assumed the density of events is constant for all the n_1 events in the path for simplicity of presentation. The changing metric must however be accommodated with Eq. (50) being replaced with a sum or approximated with integrals as Eq. (33).

We can see from our simple 1+1 dimension model that the change in density of events modifies the action along the path (Fig. 7). For Minkowski space–time the action is

$$S_1 = \sum_i^{n_1} -m_1 c^2 \Delta \tau_1 = -n_1 m_1 c^2 \Delta \tau_1 = -m_1 c^2 \tau. \tag{53}$$

For two different masses with the same velocity, for simultaneity and equivalent τ the ratio of actions is just the ratio of masses

$$\frac{S_2}{S_1} = \frac{m_2}{m_1} = \frac{\sigma_2}{\sigma_1},$$

(54)

and equal to the ratio of density of events along the path. The additional action δS due to a change in the density of events $\delta \sigma$ in the path is

$$\delta S = -m(\frac{\delta \sigma}{\sigma})c^2 \tau,$$

(55)

which we can approximate as a volume integral over an appropriate energy density ρc^2

$$\delta S = -m(\frac{\delta \sigma}{\sigma})c^2 \tau \approx -\int \rho c^2 d^3 x d\tau.$$

(56)

This provides a link to the gravitational action

$$\delta S = -\int \rho c^2 d^3 x d\tau = \int T d^4 x = -\int \frac{R}{K} d^4 x.$$

(57)

Here we have used the trace of the Einstein field equation

$$G = KT = -R,$$

(58)

and its relation to the Reimann curvature R and the constant K. The Einstein–Hilbert action [6,7] can be then written in the coordinate system $\hat{x}$ as

$$S_g = \int_{\Omega} T\sqrt{-g} d^4 \hat{x} = -\frac{1}{K} \int_{\Omega} R\sqrt{-g} d^4 \hat{x}.$$

(59)

Variation of the Einstein–Hilbert action recovers the complete Einstein field equations in a standard way [6,7]. This demonstrates that the causal net model based on possible events is compatible with Einstein's general theory of relativity.

5. Causal net cosmology

Common cosmological models have the following metric form [6,7,16,17]

$$ds^2 = c^2 dt^2 + a^2(t)d\kappa^2.$$

(60)

where for example

$$d\kappa^2 = d\chi^2 + \sin^2\chi(\sin^2\theta d\phi^2 + d\theta^2).$$

(61)

If $a(t)$ is assumed to be the radius or scale factor then the associated volume is usually taken as $V_0 = 2\pi^2 a^3$. These metric equations provide the well known Friedman equations for an ideal fluid with mass density ρ and isotropic pressure P

$$\frac{\dot{a}^2}{2} = \frac{4\pi G}{3}\rho a^2 - \frac{c^2}{2},$$

(62)

$$\ddot{a} = -\frac{4\pi G}{3}\left(\rho + \frac{3P}{c^2}\right).$$

(63)

The model assumes that the time is proportional to the proper time since the observer moves with the cosmic fluid and there is no peculiar motion of individual particles.

The causal net model provides a new interpretation to the classical Friedman model [6,7] since we must consider strict simultaneity and the role of possible events. In the causal net model the metric interval ds is taken as the proper time interval $\Delta\tau$ for simultaneity to be preserved. In the causal net model it is more appropriate to directly assume a is a direct function of τ. The radius or scale factor $a(\tau)$ is then equivalent to a diagonal metric $g_{\alpha\beta}$ with $g_{00} = -1$ and $g_{ij} = a(\tau)^2\delta_{ij}$.

Considering a simple 1+1 dimension universe we can see that the space–time interval can be written with the metric as as

$$-(c\Delta\tau)^2 = g_{00}(c\Delta\hat{t})^2 + g_{11}(\Delta\hat{x})^2. \tag{64}$$

which becomes in the causal net Friedman model

$$(c\Delta\tau)^2 = (c\Delta\hat{t})^2 - a(\tau)^2(\Delta\hat{x})^2. \tag{65}$$

where the spatial net scale is modified by the scale factor

$$\Delta x = a(\tau)\Delta\hat{x}. \tag{66}$$

Thus for increasing a, the observed net size $\Delta\hat{x}$ decreases in inverse proportion. Increasing energy density, increases the scale factor a and decreases the net size $\Delta\hat{x}$. For an observer this is equivalent to increasing the curvature of space. In the geodesic local Lorentz frame the net size remains constant at Δx. However, for an inertial observer within the region, an increase in a effectively "shrinks" space and the corresponding measured distances leading to the observation that the universe is expanding.

This causal net model provides the observed cosmic redshift. Consider two proper times such that $\tau_0 < \tau$ and $a(\tau) > a(\tau_0)$. If a light of wavelength λ_0 is emitted at τ_0 at a later time the standard ruler of an observer in the region has shrunk by a factor $\frac{1}{\omega}$ where

$$\omega = \frac{\lambda}{\lambda_0} = \frac{\Delta\hat{x}_0}{\Delta\hat{x}} = \frac{a(\tau)}{a(\tau_0)}. \tag{67}$$

This leads to a longer wavelength and a measured redshift.

The critical density and flatness problem is a current issue in cosmology [6,16,17]. Experimentally the observable universe is measured to be flat to high precision and the energy density of the universe is close to the critical density in the Friedman model. Here we shall consider the energy density for the causal net model. If l denotes the local Lorentz frame and o an observer frame then the energy density in the local Lorentz frame is the total energy E_T divided by the volume

$$\rho_l c^2 = \frac{E_T}{V_l} \propto \frac{E_T}{(\Delta x)^3}. \tag{68}$$

However in the observer frame, within the expanding region, the equivalent volume element is smaller so the volume for the observer is measured as larger, giving

$$V_o = V_l\frac{(\Delta x)^3}{(\Delta\hat{x})^3} = V_l a^3, \tag{69}$$

since $\Delta x = a\Delta\hat{x}$. This gives the relative energy densities in the two frames as

$$\rho_l c^2 = \frac{E_T}{V_l} = a^3\rho_o c^2, \tag{70}$$

$$\rho_l = a^3 \rho_o. \tag{71}$$

The energy density of the observer space–time $\rho_o c^2$ multiplied by the volume scale factor a^3 is thus equivalent to the effective energy density of a causal net built in the local Lorentz frame. The energy density of the causal net in the local Lorentz frame is equal to that of flat Minkowski space–time. The well known relation for the Hubble constant [6]

$$H^2 = \frac{8\pi G}{3}\rho - \frac{kc^2}{a^2}, \tag{72}$$

can be rewritten as

$$\rho_c a^3 - \rho a^3 = -\frac{3akc^2}{8\pi G}, \tag{73}$$

where the critical density is given as

$$\rho_c = \frac{3H^2 c^2}{8\pi G}. \tag{74}$$

For a causal net any allowable energy density ρc^2 multiplied by a given a^3 is equal to that of the flat local Lorentz frame net so

$$\rho_c a^3 = \rho_o a^3 = \rho_l, \tag{75}$$

and so $k = 0$ for the causal net model whatever the scale factor a. This provides an indication that space–time is always flat for an essentially quantum universe composed of causal nets, since the energy density times the scale factor a^3 is that of the flat geodesic causal net. The total energy E_T is given by the volume integrals

$$E_T = \int \rho_o c^2 dV_o = \int \rho_l c^2 dV_l. \tag{76}$$

Thus the total causal net of energy E_T has different energy densities in the observer and the geodesic local Lorentz frame but the volume element used in integration scales to compensate.

6. Causal net dark energy

Another outstanding problem in cosmology which we can tentatively attempt to model with quantum causal nets is that of dark matter and dark energy [6,16,17]. In the causal net model the universe is comprised of a continuum of causal nets composed of possible events. These events comprise both visible and dark matter. Dark energy is arguably due to causal nets exerting an internal quantum pressure from velocity reversals or "Zitterbewegung". For a combination of opposing causal nets, there may be no net average velocity and, although there is no pressure gradient, the internal kinetic energy contributes to the pressure.

If we assume the universe is composed of causal nets that behave as an ideal fluid with an energy density ρ and local hydrostatic pressure P. The stress energy tensor is [6]

$$T_{\mu\nu} = (\rho + P/c^2)v_\mu v_\nu + (P/c^2)g_{\mu\nu}. \tag{77}$$

Assume the average universal 4-velocity is zero and space is composed of a multitude of causal nets. Consider a single causal net representing a single free particle of mass m and a given net energy E. The energy transfers at each causal net flux vertex can be decomposed into an expected energy "flux" F of

$$F = E\hat{P}_{11} - E\hat{P}_{21} = mc^2, \tag{78}$$

and an internal "pressure" P of

$$P = 2E\hat{P}_{12} = 2E\sin^2(\theta/2), \tag{79}$$

such that these equal the energy E

$$F + P = E. \tag{80}$$

For a velocity or flux in any direction in 3 dimensional space a hydrostatic pressure is observed. In the cosmic fluid with no net average velocity there exists an internal pressure P but the net aggregate expected flux universally is zero. The pressure P is thus similar as Bernoulli's theorem to the kinetic energy or relativistically to the energy above that in the stationary local Lorentz frame mc^2 since

$$P = 2E\hat{P}_{12} = E - mc^2 = mc^2(\gamma - 1). \tag{81}$$

Consider the trace of the ideal fluid stress energy tensor

$$Tr(T) = -\rho c^2 + 3P. \tag{82}$$

This can be written alternatively as

$$Tr(T) = -E + 4P. \tag{83}$$

The ratio of dark energy to matter energy $\frac{\Omega_\Lambda}{\Omega_M}$ can be considered as the ratio of pressure to energy. For a single net angle we can write the ratio as

$$\frac{\Omega_\Lambda}{\Omega_M} = \frac{4P}{E} = 8P_{12} = 8\sin^2(\theta/2). \tag{84}$$

If we make a simplifying equipartition assumption, that the causal nets for all possible velocity states contribute to the total pressure, then we can sum over the pressure terms for all velocities or angles at each event

$$\frac{\Omega_\Lambda}{\Omega_M} \approx 8 \int_0^{\pi/2} \sin^2(\theta/2)d\theta. \tag{85}$$

This gives as an estimate for the observed matter energy to dark energy ratio

$$\frac{\Omega_M}{\Omega_\Lambda} = \frac{1}{4(\frac{\pi}{2} - 1)} \approx 0.438. \tag{86}$$

Current experimental estimates [6] for dark energy are 0.7, dark matter 0.25 and normal matter 0.05. These values would provide an estimate of the ratio $\Omega_M/\Omega_\Lambda \approx (0.25 + 0.05)/0.7 = 0.428$. This value is close to the value derived from the simple causal net approach, based on equipartition of plane wave states.

The ratio of the hydrostatic pressure energy to the observed energy is consistent with all possible causal nets existing prior to a measurement and contributing to a quantum dark energy. If an event or particle is measured, the quantum possibilities prior to the measurement manifest as a pressure or dark energy produced by the quantum "Zitterbewegung" of the different causal trajectories in space–time.

The above analysis for dark energy assumes the existence and summation across all causal nets with the velocity in one defined space direction. However, for each causal net vertex of a given energy or net angle there are 6 possible orthogonal velocity basis directions. A general solution must consider a superposition or sum over these states. Only one velocity direction can constitute an observed or measured trajectory whilst the other 5 basis directions will remain unobserved. If these unseen quantum trajectories constitute equally dark matter then the ratio of observed matter to dark matter would be approximately 1:5. This would be in broad agreement with experimental observation [6]. The dark matter then possibly originates from the hidden quantum states, which contribute to the overall density of possible events in space–time. The gravitation of quantum objects would thus appear different from that of classical matter. If visible, classical matter is composed of measurable, well-defined trajectories in space–time, then dark matter is perhaps the ephemeral, possible events of causal nets that do not observationally manifest themselves as long-lived particles.

7. Discussion

The causal net approach to relativistic quantum mechanics [1] where space–time is built from possible point events can be extended to curved space–time to provide an analog to gravitation. Inertial mass in the causal net model is provided by the net scaling or density of events in space–time. The density of events has a similar relation to the spatial net spacing or wavelength as a traditional refractive index $\frac{n_1}{n_2} = \frac{m_1}{m_2} = \frac{\lambda_2}{\lambda_1}$. The density of events in space–time enters into the metric for the causal net, even for the Minkowski flat metric, which suggests an equivalence between inertial and gravitational mass.

For curved space–time the paths of the causal net are geodesics and the Dirac equation in the local Lorentz frame is unchanged in these geodesic coordinates. In the vicinity of an event space–time is locally flat and the causal net angle and probabilities linking events are preserved. The curvature of space–time in the causal net model is produced by variation of the density of possible events in the causal path. If relativistic simultaneity is maintained, this leads to changes in the metric. The density of events modifies the action and provides a route to the Einstein–Hilbert action and general relativity.

The overlap and interaction of events from the causal nets representing different particles can alter the density of events and is the origin of an effective gravitation in the causal net model. In a simplistic sense, if a particle with mass m is proportional to a number of events n and a second particle mass M a number of events N, then if each event has a causal relation or linkage to that of the other particle then the number of causal relations is nN and is proportional to mM. This product of masses is similar to gravitational forces and potentials in Newtonian gravitation.

The causal net model can be applied to cosmological models of the universe. The Friedman equations assume a scale factor a for cosmic expansion and the causal net model with its rigorous application of simultaneity provides a different interpretation for cosmic evolution. In a space–time region where a increases the net size decreases so that for an observer within the region the universe is seen to expand whereas, for an local Lorentz frame observer the region retains the same volume. In this way the total energy of the system is conserved and remains the same as that of the causal net in the local Lorentz frame. Thus in the causal net model the universe energy density is the critical value for flatness derived from the local Lorentz frame.

A possible solution to dark energy is provided by a quantum pressure produced by all the possible causal paths existing prior to a measurement. The effective pressure from each causal net is a result of "Zitterbewegung" and proportional to the reversal probability at each vertex of the causal net. For a cosmological fluid model the equilibrium ratio of matter energy to hydrostatic pressure energy is similar to that observed experimentally for the ratio of matter energy to dark energy. Dark matter potentially results from the combined density of unobserved events from different orthogonal spatial directions at each net vertex. In the causal net model the gravitation of quantum objects would differ from classical matter in exhibiting both dark matter and dark energy.

References

[1] Bateson R D 2012 *J. Phys.: Conf. Ser.* **361** 012009
[2] Dirac P A M. 1928 *Proc. Royal Soc.* A 117
[3] Feynman R P and Hibbs A R 1965 *Quantum Mechanics and Path Integrals* (McGraw-Hill)
[4] Dirac P A M 1951 *Nature* **168** 906
[5] Dirac P A M 1951 *Proc. Royal Society A* **209** 291
[6] Carroll S M 2019 *Spacetime and Geometry* (Cambridge)
[7] Misner C W, Thorne K S and Wheeler J A 2017 *Gravitation* (Princeton)
[8] Rindler W 1982 *Introduction to Special Relativity* (Oxford)
[9] De Broglie L 1925 *Ann. Phys.* **3** 22
[10] Heisenberg W 1927 *Z. Phys.* **43** 172
[11] Schurmann T and Hoffmann I 2009 *Found. Phys.* **39** 958
[12] Davydov A S 1965 *Quantum Mechanics* (Pergamon Press)
[13] Coulter B and Adler C 1971 *AJP* **39** 305
[14] Li C and Ma B 2022 *Phys. Lett. B* **829** 137034
[15] Feynman R P, Morinigo F B, Wagner W G 2003 *Feynman Lectures on Gravitation* (ABP)
[16] Zel'dovich Y B 1968 *Usp. Fiz. Nauk* **95** 209
[17] Mukhanov V 2005 *Physical Foundations of Cosmology* (Cambridge)